Morphogenesis of Industrial Symbiotic Networks

Erika Džajić Uršič

Morphogenesis of Industrial Symbiotic Networks

PETER LANG

**Bibliographic Information published by the
Deutsche Nationalbibliothek**
The Deutsche Nationalbibliothek lists this publication in the Deutsche
Nationalbibliografie; detailed bibliographic data is available online at
http://dnb.d-nb.de.

Library of Congress Cataloging-in-Publication Data
A CIP catalog record for this book has been applied for at the
Library of Congress.

"This book was created in cooperation with the School of Advanced Social
Studies in Nova Gorica and the Faculty of Information Studies in Novo
mesto, Slovenia."

ISBN 978-3-631-80200-7 (Print)
E-ISBN 978-3-631-80659-3 (E-PDF)
E-ISBN 978-3-631-80660-9 (EPUB)
E-ISBN 978-3-631-80661-6 (MOBI)
DOI 10.3726/b16330

© Peter Lang GmbH
Internationaler Verlag der Wissenschaften
Berlin 2020
All rights reserved.

Peter Lang – Berlin · Bern · Bruxelles · New York · Oxford · Warszawa · Wien

This publication has been peer reviewed.

www.peterlang.com

Abstract: The present work deals with Industrial Symbiosis (IS), a term used to describe a network of diverse organizations that make use of different by-products to improve their facility to achieve common goals, improve environmental conditions or improve business and technical processes. Industrial symbiosis is understood as a technological material as well as a social relationship between so-called social actors, which are involved in the exchange of secondary resources.It is relevant to deal with the paradigm of industrial symbiosis, more precisely with the knowledge that industrial symbiosis can take place within one or several Industrial Symbiotic Networks (ISNs), which are involved in mutual cooperation, because although participation in industrial symbiosis seems rational both from an environmental, organizational, and societal point of view in the scenario that if we had sufficient technological and logistical knowledge to manage them successfully, then industrial symbiosis could be further developed both on regional, national, and international levels (in forming new industrial symbiotic relationships and networks).The author proposes a model for the evaluation of the possibilities to establish such industrial symbiosis with a study benchmark of seven industrial symbiotic examples used to build a qualitative multi-criteria decision model for the evaluation of the industrial symbiotic network model. With the data obtained from the best-known industrial symbiotic cases in the world, the author examines the importance of social actors' involvement in industrial symbiosis both in their industrial and non-industrial technological processes. At the same time, the author demonstrates an impact of social forces on the development of industrial symbiotic networks (speaking of networks, cognitive frames, and infrastructures) in which connections among firms are not only technological networks, but are also social networks. The author demonstrates that all the people and firms involved in industrial symbiosis follow their own benefits and, at the same time, they provide different benefits to others involved and society in general, which influences the social involvement of each involved in industrial symbiosis. The study not only presents a methodology, the qualitative multi-criteria decision model, but also provides a support to decision makers in their decisions to undertake future regional industrial symbiotic projects or constructions. This study can be a good design for symbiotic initiators, who must adapt the procedures to their industrial, economic, environmental, and social contexts.

Keywords: Industrial symbiotic networks, social forces, modelling, multi-attribute hierarchical model, evaluation

Table of contents

1 Introduction

The important nature of cost productivity, environmental safeguard, resource management, greenhouse gas emissions/reduction, and social considerations are imposing increasing pressure on the industrialized districts. Industrial symbiosis emerged as a collective, cooperative, and multi-industrial approach towards economic and environmental sustainability. Industrial symbiosis has been defined as bringing together »traditionally separate industries in a collective approach to competitive advantage involving the physical exchange of materials, energy, water and by-products« using Chertow's definition (2000, 313–337). The positive economic effects are demonstrated as lower waste/by-product disposal costs, competitive advantages towards other firms, and a better reputation of the related firm that is also social actor. Firms hence participate in industrial symbiosis for the exchange of resources, such as by-products, materials, services, energy, etc., in the framework of closed and sustainable cycles. This cooperation contributes to the reduction of required input of materials and energy, as well as a decrease of the output in the form of waste. Furthermore, the low price of waste as a material, if compared with no waste materials, increases the economic effectiveness of individual organizations, which are recipients in the framework of industrial symbiotic networks. Open networks, on the other hand, allow entropy, which results in a loss of energy, poor utilization of materials, and consequently causes environmental damage (Chertow 2000, 2007).

The current study will analyse networks as parts of social fields in the industrial symbiosis, in which these networks help shape their topography. This will also have a more profound scientific effect, as the social topography of industrial symbiosis will serve as a case study of interaction among individual actors with their environment, based on their specific needs and preferences. Why is this the case in a present study's research? Researching industrial symbiosis, combining its technological, ecological, informational, economical, and, as of recently, sociological aspects, can be relevant to study and to put forward a simple method for creating a model of industrial symbiotic networks. Keeping in mind that social interaction plays a major role in industrial symbiosis, including on a micro-mezzo-macro level, social structures are best suited to show the emergence of patterns from individual decisions and social interactions. By introducing a new model, this study explains and possibly forecasts new industrial symbiotic networks and their »behaviours-interactions« in a new region. By analysing such a model, it would be possible to identify some critical factors that should be considered when studying social interaction and their role for explaining observed patterns (Džajić and Rončević 2017).

The study of industrial symbiotic networks from the economic and especially ecological point of view is currently becoming the start of the subject of numerous multidisciplinary (and interdisciplinary) research studies, which mainly focus on waste management (see Boshkoska, Rončević and Džajić 2018, CTTÉI 2013), industrial ecosystem (Chertow 2000; Ehrenfeld and Gertler 1997), rounded economy (Laybourn 2016), supply chain management (Leigh and Li 2015), and economics (Chertow and Lombardi 2005; Džajić and Rončević 2017 quoted in Tomšič and Rončević 2017; Modic and Rončević 2018).

1.1 Objectives and the purpose of this study

This study will deal with the development of industrial symbiotic networks, where industrial symbiotic networks are dynamic, evolving, complex networks that are transcending the boundaries of technical, natural computer, and social systems.

The objectives and aims of the proposed study are as follows:

- To show connections and exchanges between dynamic, completive, adaptive networks, operating across the boundaries of technical, natural, computational/informational, and social systems.
- The development of a robust input-output model of physical, organizational, and technological exchanges, including their social interactions (networking); the ultimate aim is to model morphogenesis of industrial symbiotic networks by taking all relevant aspects into account. This will be engaged in computational sociology in order to explore, model, simulate, test, and observe an industrial symbiotic network.
- To allow us to predict morphogenesis in a specific social setting using the model, thereby allowing us to develop tools for embedding industrial symbiosis in its natural, technical, and sociocultural setting.
- To use the model in order to be able to identify the prerequisites and mechanisms for successful morphogenesis of industrial symbiotic networks, which will enable us to highlight already existing potentials and barriers to industrial symbiosis within a specific setting. This includes factors which are necessary to improve collaboration amongst actors.
- To use the research to enlighten social aspects of industrial symbiotic networks and to use these theories to conceptualize the morphogenesis of symbiotic networks; it will also examine the influence of social structures on individual actions of industrial symbiotic networks development.

The purpose of this study is to improve the methodology of the development of industrial symbiotic networks in industrial symbiosis, which is composed of various data analysis techniques. The new model allows for the evaluation of the current state of the industrial symbiotic network based on the three spheres and contributes to its clarification and to finding weak points in order to improve the industrial symbiotic networks.

1.2 Research questions and hypothesis

The research will be able to explore not only the explanatory role of social theory, but also its applied potential, the theoretically inspiring morphogenesis of industrial symbiotic networks. Exploration will be of one of the basic sociological questions, namely the mutual influence of emergent social constructions and individual actions. Next, the study and explanation of substantial differences in the morphogenesis of industrial symbiotic networks in different sociocultural settings and with the development of information and communication technologies, numerous formal and informal networks start to emerge.

Organized networks of structures and organized networks of actors are referred to as collaborative networks of industrial symbiosis (Archer 2003). In industry, clusters of organizations aim for efficient product development, production, and marketing tasks. Such partner consortia function as networked organizations in which cooperation is supported by information and communication technologies. Mutual trust can contribute to improved knowledge sharing, resource sharing, and taking joint risks. Trust brings many advantages to industrial symbiotic networks and should, therefore, be considered beforehand when creating industrial symbiotic networks. The branch of metaphysics (the philosophy concerning the overall nature of what things are) is concerned with identifying, in the most general terms, the kinds of things that exist in industrial symbiotic networks. In other words, addressing the following research questions:

- What are the chances that industrial symbiotic networks can materialize?
- Despite social theories, how do social systems grow up?
- How can the industrial symbiotic be integrated into a society where there is not yet such a system?
- What is the definition of existence, and what is the nature of existence of industrial symbiotic networks?
- Is the branch of philosophy concerned with the nature of knowledge itself, its possibility, scope, and general basis, too broad?

- How do we begin to know things regarding industrial symbiotic networks?
- How do we separate true ideas from false ideas in industrial symbiotic networks?
- What are the systematic ways in which we can determine when something is good or bad in industrial symbiotic networks?

Therefore, ontological and epistemological premises guide not only the selection of the research focus, but also the methodology adopted and the outcomes expected. Industrial symbiotic networks are thus examined as a social structure that is simultaneously constructed and reflected upon by the researcher and actors involved (Boshkoska, Džajić and Rončević 2015; Adam, Makarovič, Rončević and Tomšič 2005). This approach understands their identification first as a singular phenomenon, and then as a sustainable instrument which is part of the process of the co-creation of meaning and social interaction. This study focuses on the social processes behind the articulation of industrial symbiotic networks and the emphasis given to structural and discursive dimensions in the analysis.

Social networks are more complex, encompassing tens of thousands of actors (or more) with many possible topological and structural patterns (Watts 1999, 493–527, 2004). While answering research question, we will test the following *Hypothesis:*

> »*Industrial symbiosis is not only a paradigm with important technological, economic, and environmental consequences. Industrial symbiosis is more than environmentally sustainable industrial activity. Industrial symbiosis is also a social paradigm and its manifestations in the form of social networks are an indistinguishable part of the development process of Industrial Symbiotic Networks.*«

1.3 The structure of the work

The main aims, objectives, research questions, and hypothesis are presented in Chapter 1. Chapter 2 provides a literature overview of the concept theory, development, tools, and agenda of industrial ecology and industrial symbiosis. Also provided is a characterization of the existing literature on industrial ecology and industrial symbiosis, assessing the concepts and theoretical insights that have been proposed to aid in the understanding of the phenomenon industrial symbiosis. It describes how industrial symbiosis are networks of firms that engage in mutually beneficial waste, energy, material, and knowledge exchanges. This chapter reviews the main analysis of the state-of-the-art study on the history and development of the industrial symbiosis and outlines its relation with other partially overlapping concepts such as industrial clusters or industrial districts, which can be the site for the formation of industrial symbiotic networks. The eco-industrial paradigm, its role in the process of green industry and progressing towards more sustainable industrial systems are presented. This will include, but not exclusively, a critical assessment of economic environmental performance brought about by industrial symbiosis as many eco-industrial parks have been identified across all continents. From the literature of eco-industrial development, some potential opportunities and obstacles that could challenge the further implementation of this approach are identified.

In Chapter 3, the knowledge of the concept of sustainable development in industrial ecology further explains the following question: What kind of modern economy should exist so that we can achieve the sustainability of humanity? Some core features of industrial symbiotic networking, which contribute to this study research and this study field, pay special attention to the social dimension of these networks. Chapter 3 outlines social factors in developing social links in the industrial symbiosis from the analysis of the embeddedness, social mechanisms, and social field theory. Considering why, under similar material conditions, the actual emergence and operation of social networks can result in radically different outcomes, depending on the social-institutional structure in which they operate, allows a hypothesis to be tested concerning the relevance of the socioinstitutional scenario in the development of industrial symbiotic networks (beyond the merely material and technical conditions). From a comparative analysis, some structural features of successful networks can be identified in the next chapter.

Chapter 4 relies on the cross-comparison of multiple case studies. Different typologies of industrial symbiosis have been selected and described as case studies: Kalundborg, Aalborg, Guayama, Barceloneta, Gladstone, Kwinana, and Rotterdam. From the author's point of view, all of those cases have been chosen as »efficacious and operational« examples of industrial symbiosis. The most recognized Kalundborg industrial case was selected as the paradigmatic case. The contribution of these cases is precisely to point out the socioinstitutional elements that may hinder the full development of industrial symbiotic relations, even in contexts where there is a combination of activities that provide an ideal material basis for industrial symbiotic exchanges and networking. A proposed morphology assessed model in this chapter shows some recommendations for the implementation of the advanced data analysis techniques with a software tool.

In Chapter 5, the research questions and hypothesis are defined and the general study design strategy is outlined. In addition, the data collection methods and general methodological approach is explained. There is also a discussion about the main caveats and limitations of the methodology approach adopted. All of the included descriptions of the method applied are based on a software tool and a decision expert method.

Chapter 6 analyses and evaluates the degree of compliance of development industrial symbiotic networks with the industrial symbiotic prerequisites (attributes) of seven industrial symbiotic networks. Once the general methodology of this study has been outlined, this chapter focuses on the specific analytic structure adopted to analyse the data and generate the theory. It will then be allowed to determine what are the facilitating mechanisms already existing, what are the barriers to avoid, but also to give recommendations regarding future operations to be undertaken.

Chapter 7 provides a discussion and summarizes answers to the research questions, defines the results and study hypothesis, based on the research study of industrial symbiotic cases and evaluation results. In the appendix, are detailed results-outputs from used software tool. All literature, sources, and references used throughout this study are listed at the end of the book.

2 Literature review

During the last two decades, increasingly pushing constraints of an economic and environmental nature have appeared. Global warming, soil impoverishment, etc., are worrying facts that forced governments to pledge positions to limit human impact on the environment. The most notable change was undoubtedly the implementation of management systems within organizations, aiming at reaching a sustainable situation. The outputs of each integrate firm becomes the inputs of another. The integrate firms are in a network of services, by-products, and energy flows. That would make them dependent on each other. The firms can also share ancillary services, such as transportation, landscaping, and waste collection, and share in the management of their utilities, such as energy, water, or waste-water treatment. Every day we hear different reports on increasing air pollution, the environment and the pessimistic predictions of scientists regarding the expected state of the environment, and the problem of oversized populations in the future. We are aware of all that. We have everything we need; the impact on the environment, however, is not such a burning topic. Is this exact? To be able to understand why it is necessary to invest in ecology and include it in our daily lives, we must first know the concept of sustainable development, which is introduced in the first part of this study. This is the theoretical framework that defines objectives, which must be achieved, so that children can live decently, at least as good a life as the previous generation. To achieve a better living environment in the future, we have to give up a specific truth in the present. It comes in changing human habits and the renunciation of certain comforts, so we firstly need to be precisely aware of the importance of sustainable development and then conceive higher goals.

The worldwide industrial activity has not ceased to increase since the industrial revolution of the 19th century. Thanks to technical progress, workers have become more productive. We now produce more with less. Up until forty years ago, the world did not seem to have limits, and therefore, could produce and provide with resources ad infinitival. Industry acts as a major player in energy consumption and accounts for approximately 1/3 of consumed world energy (International Energy Agency 2008). A sub-consequence of the massive use of fossil energies is global warming, caused by the emission of greenhouse gases that retain the sun's heat within the atmosphere. Once again, the industry sector, alongside the freight industry, shares the main responsibility for that. It would directly account for 17 % of the total greenhouse gas emission and indirectly for

30 % of the transformation of primary energy into final energy for industrial use (Climate Change Synthesis Report 2007). In addition, according to the European Commission for statistics, the overall volume of waste generated originates from the following economic sectors: agriculture, industry, construction, and services. In 2008, industry- and construction-generated waste volume peaked up to 90.7 % of all waste produced by human activities (European Commission Eurostat 2010; Statistics Explained n.d.).

Governments undertook initiatives leading to a sustainable management of the available and proven energy resources. The United Nations Environmental Program (UNEP) – dealing among others with solid waste and sewage management issues – recommends in the agenda 21 program that »waste minimization technologies and procedures will need to be identified and widely disseminated« (Report of the United Nations Environment Programme 2017).

2.1 A circular economy as economy of the future

One of the bases that provide a recuperative system where materials flow in a stable system, providing balance between economic development and ecological resource protection, is »circular economy«. If outcomes are advantageous to not only to the firm, but to the environment as well, then ecological economic principles are essential (Behne 2016).

This more sustainable approach, where efficient distribution of resources and a philosophy of an end to development, drives this single economy to be different – »circular-economy«. Koenig (2009) explains that »circular economy« is a rounded or complete economic concept. It seeks efficiency in resources used through the combination of cleaner production and industrial ecology into a broader system surrounding industrial firms, eco-industrial parks, networks, chains of firms, and regional infrastructures to support resource optimization. Government-owned and private firms, government and private infrastructure, and customers, all create the circular economy (Behne 2016).

Circular economy returns any materials back to environment compassionately, but industrial ecology grows even more *difficult ideas and truths*. A very noted example from the circular economy, or later known as industrial symbiotic thoughtfulness, can be seen in the development of businesses in Aalborg in recent years. There, they are experimenting with industrial symbiotic initiatives seen in several projects, where the target is to foster the collaboration in-between the firms in sharing their wastes, resources to minimize environmental impact, and to share knowledge (van Bosch-Ohlenschlager 2010; Kerndrup, Søren, and Xavier Gabarrell Durany 2014).

2.2 Introduction to industrial ecology concept

Industrial ecology, which is sometimes described as »an integrated system, in which the consumption of energy and materials is optimized and the effluents of one process serve as the raw material(s) or energy for another process« (Frosch and Gallopoulos, 1989, 144; Baas and Huisingh 2008, 399–421), is an increasingly well-known concept, worldwide.

> As argued in the seminal publication by Frosch and Gallopoulos (1989) that did much to coalesce this field, industrial ecology looks to non-human »natural« ecosystems as models for industrial activity«. Firstly, this is what some researchers have dubbed the »biological analogy« (Wernick and Ausubel 1997; Ayres and Ayres 2002; Allenby and Cooper 1994). Many biological ecosystems are especially effective at recycling resources and thus are held out as exemplars for the efficient cycling of materials and energy in industry. Secondly, industrial ecology places human technological activity-industry in the widest sense-in the context of the larger ecosystems that support it, examining the sources of resources used in society and the sinks that may act to absorb or detoxify wastes. This latter sense of »ecological« links industrial ecology to questions of carrying capacity and ecological resilience, asking whether, how and to what degree technological society is perturbing or undermining the ecosystems that provide critical services to humanity (Suh 2009, 4).

More simply economic systems are viewed, not in separation from their nearby systems, but in performance with them (Lifset and Graedel 2002, 3–15). The industrial ecology concept was introduced to industrial managers of firms as a paradigm of prevention, which was oriented for achieving cleaner industry and more sustainable societies (Lifset and Graedel 2002).

The complexity and uncertainties of testing and implementing new concepts, procedures, and technologies are often approached from the perspective of scepticism, ignorance, misperception, and the fear of making any changes (Commission on Sustainable Development 2005; Baas and Huisingh 2008, 399–421). In this context, it is increasingly being found that previously, there had been too much emphasis upon the technological and mechanical dimensions of change and far too little emphasis upon understanding and working with the non-technical dimensions. Therefore, better success is being achieved by the integration of the economic, environmental, and social dimensions into the industrial ecology activities. In fact, this integration is being documented to be an essential condition for making progress towards a more sustainable society (Baas 2008, 330–340; Baas 2005).

Industrial ecology has sprung up all over the world, mostly in regions with large-scale, heavy industrial processes, but also around small-scale agricultural practices, such as the Montfort Boys Town in Fiji (Ashton 2008). In the

USA, though more than a dozen eco-industrial developments were planned, none materialized as idealized industrial ecologies (Chertow 2007; Gibbs and Dutz 2005, 452–464). Many nascent industrial ecologies are thought to exist, although they are not yet known to industrial ecologists. These include regions with by-product exchanges amongst small groups of firms or utility-sharing agreements to deal with local resource deficiencies (Chertow 2007).

In the last decade, global trends in environmentally related issues within Multinational Corporations (MNCs) have revealed that more and more of them are gradually incorporating different dimensions into their policies and operational strategies. The progression is now to incorporate »Corporate Social Responsibility« (CSR) into their approaches (ecology, economy, and social aspects). Involving different firms and actors (including their different activities and targets) that are required for the existence and development of industrial ecology in a region is nowadays an important, but time-consuming structure (Baas 2005). As in all sectors of the exchange of goods and services, the economy and economic relations play a major role. According to Granovetter (1985, 481–510), economic relations between individuals or amongst firms are embedded in actual social networks and do not exist in an abstract idealized market.

Experiences founded in industrial ecology projects around the world show that an open, reflective, and ongoing dialogue must be designed to develop *trust and transparency* to ensure the real involvement of diverse actors in charting the future of their organizations and regions as part of the transition to sustainable societies. Consequently, new types of abilities, skills, knowledge, and social behaviour are needed to support changes beyond the status quo situation (Baas and Boons 2006).

2.3 Introduction to industrial symbiotic concept

The concept of industrial symbiosis is closely linked to that of »Eco-Industrial Parks« (EIPs). But how these two concepts of industrial symbiosis and EIPs relate exactly is still a matter of debate. Gibbs and Deutz (2005, 452–464) state that it is unclear whether by-product exchange should be seen as a defining feature of an EIP, and this results in the ambiguous and sometimes inconsistent use of both terms by different authors. Some authors see EIPs as one possible way of creating industrial symbiosis; in this case, it could also be reached in other ways, for instance, by a product chain approach, where firms are not necessarily located in each other's proximity (Chertow 2007). Others see industrial symbiosis as one possible configuration of EIPs; here

»eco-industrial park« is interpreted more broadly as an industrial park where firms work together to achieve sustainability goals, not necessarily involving by-product exchange (Thongplew 2015, Thongplew, Spaargaren, and van Koppen 2014, 99–110).

However, it is important to note that, because of the strong interconnected (and sometimes directly interchangeable) nature of the two concepts, we consider many of the observations made in literature about EIPs to be relevant for our current investigation of industrial symbiosis.

Different authors define industrial symbiosis in varying ways, such as an approach to industrial ecology; a synonym for industrial ecology (Phillips et al. 2006, 242–264); a subset of industrial ecology (Chertow et al. 2005, 6535–6541); an activity in industrial ecology (Rui and Heijungs 2010); an eco-industrial symbiosis which represents local or regional circular economy and environmental approach (Chertow 2007; Howard-Grenville and Paquin 2008, 157–175; Gingrich 2012, 44–49; Hartard 2008). Deutz (2014, 3–13) sees industrial symbiosis as a stream of available resources (including objects, energy, materials), starting with social players, firms that would have discarded these resources, and ending with social actors, that use these resources as new resources. Some are also attempting to be recognized as environmentally aware by their social environment (Costa and Ferrão 2010, 984–992; Costa et al. 2010, 815–822; Eilering and Vermeulen 2004, 245–270). Both economic considerations are motivated by such aspects as lowering the costs for waste disposal, in addition to environmental motivations, such as accessing limited water supplies. Firms operating within industrial ecology paradigm use industrial symbiosis, for example, a collective approach to joint competitive advantage, usually within the framework of a geographically delimited cluster or industrial district, simultaneously realizes environmental benefits. Through such collaboration, social relationships among participants improve (Zhu et al. 2007, 31–42).

Without a doubt, industrial symbiosis is the most concrete example of the concept of industrial ecology. The challenge lies in circulating one's residual materials and energy to substitute the inputs of another. The objective is thus to prolong the life cycle of resources by substituting and mutualizing material flows. This circular movement also applies to human resources through expertise and services exchanges, equipment sharing, etc. (National Zero Waste Council Circular Economy 2015; Synergiequebec 2017; Fric and Rončević 2018).

As previously seen, authors differently describe and conceptualize the connection amongst the flows of services, resources, by-products, waste, and an emerging field of industrial ecology. It demands resolute attention to the flow of materials and energy through local, regional, and global economies. The concept of industrial ecology so »requires that an industrial system be viewed not in isolation from its surrounding systems, but in concert with them. It is a system view in which one seeks to optimize the total materials cycle, from virgin materials, to finished material, to component, to product, to obsolete product, and to ultimate disposal« (Graedel 1996, 70). Factors to be optimized include resources, energy, and investments (Ayres and Ayres 2002; Graedel and Allenby 2003). Therefore, industrial ecology allows attention at the facility level, at the inter-firm level, and at the regional or global level. On the other side, industrial symbiosis occurs at the inter-firm level because it includes exchange options among several organizations.

> The expression »symbiosis« builds »on the notion of biological symbiosis relationships in nature, in which at least two otherwise unrelated species exchange materials, energy, or information in a mutually beneficial manner-the specific type of symbiosis known as mutual cooperation. So, industrial symbiosis consists of place-based exchanges among different entities. By working together, businesses strive for a collective benefit greater than the sum of individual benefits that could be achieved by acting alone. This type of collaboration can advance social relationships among the participants, which can also extend to the surrounding neighbourhood. As described, the symbioses need not occur within the strict boundaries of a »park«, despite the popular usage of the term Eco-Industrial Park to describe organizations engaging in exchanges (Chetrow 2002, 314).

While interest began to develop in industrial symbiosis and eco-industrial parks, several other parallel tracks advanced that might be construed, broadly, as »green development«. These include sustainable architecture, green building, sustainable communities, and smart growth, among many other terms. In the Rocky Mountain Institute's Green Development: Integrating Ecology and Real Estate, the authors point out that there is no single face to this kind of firm because »for one project, the most visible »green« feature might be energy performance; for another, the restoration of the prairie ecosystems; for yet another, the fostering of community cohesion and reduced dependence on the auto mobile« (Institute Rocky Mountain 1998; Chertow 2000, 313–37).

The industrial symbiotic approach in Kalundborg, Denmark, is often used as an illustration of outstanding developing linkages of utilities and waste applications amongst firms in an industrial complex (Ehrenfeld and Gertler 1997; Gertler 1995).

Despite that and other successes, the widespread adoption and implementation of the concepts and practices presented in this study related to industrial ecology or industrial symbiosis have been found to be a difficult and slow process.

2.4 Barriers and obstacles of industrial symbiosis

As Andrews (2000 quoted in Ayres and Ayres 2002, 14) *discuss: one of the considerations with respect to the boundaries is whether industrial symbiosis (and industrial ecology) should address the question what, but also how«. Analyses of what tell to our understanding of the technological and natural systems' character, characterizing the way these systems behave and interact, and under what conditions environmental situations that humans think/believe better. For example, the absence of ozone holes in the atmosphere might happen* (Ayres and Ayres 2002, 14).

In that respect, the »what« analyses include the what-if questions described above with respect to the normative aspects of industrial symbiosis: »What if different materials were used for packaging or would carbon dioxide emissions decrease and global warming slow down?« Some industrial ecologists, however, argue that the field must also embrace social, political, and economic questions of how. That is, given the identification of a preferred outcome, what strategies should be employed to bring about that outcome (Andrews 2000; Jackson and Clift 1998, 3–5 quoted in Ayres and Ayres 2002, 14). *Thus, the how questions are largely a province of the social sciences. Sociology, Economics, Anthropology, Psychology, Political Science, and related fields have the potential to help identify strategies that are more likely to succeed. Nonetheless, the social sciences are not confined to how questions* (Fischoff and Small 1999, 4–7; Henrion and Fischoff 1986, 791–798 quoted in Ayres and Ayres 2002, 14). *They can also indicate what is happening as when, for example, social scientists investigate the quantity and character of consumption in households and how it drives production and waste management activities and therefore environmental outcomes* (Duchin 1998; Noorman and Schoot Uiterkamp 2014). *Most industrial ecologists would agree that such knowledge is crucial, but some would argue that that knowledge should remain lodged in allied fields, otherwise the boundaries and identity of industrial symbiosis (and industrial ecology) will become so expansive as to be diffuse* (quoted in Ayres and Ayres 2002, 14).

2.5 Background of industrial symbiosis as a scientific field

Industrial symbiosis has gained increasing attention, especially since 1989 when industrial ecology created a new perspective on industrial development: industrial complexes should be designed to resemble natural ecosystems to use energy, water, and material resources optimally while minimizing waste (Heeres et al. 2004, 985–995). The complexity of the process has increased in the period. Even though the exchanges of waste were not called industrial symbiotic-synergies, the early process began at that time. Among many examples, the National Industrial Symbiosis Research Programme (NISP) and the European Union associated with local partners to start up the biggest industrial symbiotic system in Tianjin Binhai New Area (China) with 80 different synergies in March 2010.

Tab. 2.1: The history of the development of industrial symbiosis. Source: Christensen
(2004); Almasi et al. (2011, 9)

1947: The term was first cited in the text of economic geography by Renner G. T. in his
book »Geography of Industrial Localization«. He describes the relationships between
organically dissimilar industries and includes the terms of »use the waste products from
one as input to another« (quoted in Almasi et al. 2011, 9).

1950: Large process industries including nickel, oil alumina refining, chemical and cement
manufacturing, and energy cogeneration plants located in Industrial Area of Western
Australia-Kwinana.

1959: In Kalundborg, Denmark started major facilities – Asnaes power plant, Statoil
refinery, and Novo Pharmaceutical plant.

1970: In Kalundborg activities began.

1989: The inter-firm linkages in Kalundborg were »uncovered« through a high school
science project, and the term »industrial symbiosis« was created to describe the system;
Frosh and Gallopoulos published the article »Strategic for Manufacturing« which is
regarded as the beginning of the field of industrial ecology.

1990: The US President's Council for Sustainable Development promoted the concept and
development of »eco-industrial parks« modelled after Kalundborg's successful inter-firm
synergies. Despite these efforts, few eco-industrial parks ever came into existence.

1996: Kalundborg Centre for industrial symbiosis was formed to facilitate inter-firms'
interactions and provide education about the system.

2000: Symbiosis activities continue to operate with new links formed between existing
entities, new facilities located to utilize by-products. The links that were no longer
economically feasible were stopped.

2001: The International Associate for industrial ecology was formed and promoted the use
of industrial ecology in education, policy, community development, and research around
the world.

2004: The First international Symbiosis Research Symposium held at Yale brought together
researchers who had practised the system from around the world; then the Symposium
was held in Stockholm, Sweden; Birmingham, England; and Toronto, Canada.

2005's NISP was launched as the first industrial symbiotic national scale initiative in the
world to promote inter-firm synergies in the regions of the UK.

---

*The literature on these industrial type of collaboration districts dates back 100 years to
the economist Alfred Marshall, who examined them to understand Britain's leadership in
textile production. A recent doctoral dissertation brought some of these topics together to
discuss how industrial collocation and inter-firm networking could lead to significant economies in environmental management related to infrastructure, information flows and regulatory enforcement, as well as to decreased conflict over land use (Kassinis 1997).*

*In international development, the term industrial estate is used to describe »a large tract of
land, sub-divided and developed for the use of several firms simultaneously, distinguished*

by its shareable infrastructure and proximity of firms« (Peddle 1993, 107–124). The United Nations (U.N.) Environment Programme issued a study in 1997 on environmental management of industrial estates. Although managing industrial estates in an environmentally sound manner is different from industrial symbiosis and describes how industrial estates are excellent places to apply the principles of industrial symbiosis because the estates contain diverse industries and can achieve economies of scale (United Nations Environment Programme 1997).

The underlying concept of industrial symbiosis is the metaphor of an industrial ecosystem that mimics a natural ecosystem, which appears early in the industrial ecology literature. In 1989, Frosch and Gallopoulos (1989) in their book inspired much of the industrial ecology that was to come when they wrote about »an industrial ecosystem« in which »the consumption of energy and materials is optimized and the effluents of one process...serve as the raw material for another process« Frosch and Gallopoulos (1989, 144). The same year Ayres wrote about both the biosphere and the industrial economy »as systems for the transformation of materials« and how studying this »industrial metabolism« could lead to shifts in the direction of increased efficiency in materials flows and waste (Ayres 1989, 23–49). In fact, in the inaugural issue of the Journal of industrial ecology, editor in-chief Reid Lifset commented on the symbiotic exchange of materials and the excitement over Kalundborg, but reassured readers that the Journal »is not simply about co-located facilities exchanging wastes« (Lifset 1997, 1 quoted in Chertow 2000, 317–318).

Table 2.1 indicates the development of industrial symbiosis in an international perspective from the late 1940s until now.

3 Outlining research of social aspects

The analysis of existing topographies is generated from several industrial symbioses cases around the world. The construction of this fact has been possible by following an analysis with a method to »guess« cause-effect relationships, in regards to a critically realistic perspective.

The metaphysical world of industrial symbiotic networks is embedded in the qualitative aspect. That was followed by observations through select literature by network theories, published interviews, and articles. With the examples of industrial symbiosis in this study, the ideology of industrial ecology was also collected, where the physical world of industrial symbiosis was built into a quantitative aspect of industrial symbiosis. Also, we can find a lot of published mass flow analysis and framework economic analysis about industrial symbiosis.

Due to the early stage development of the industrial ecology and later industrial symbiotic science, there is not much available in terms of literature, sources, and theories on how industrial symbiotic networks are developed and built. There is, yet, not enough literature on social aspects of this phenomena. The best way to obtain reliable and complete data would be to collect evidence and facts from the real world and by using open field observations. Adopted is a post-positivist critical realistic approach that allows a great range of observations (Doménech 2010).

In this study, the outlined social aspects in industrial symbiosis are not only technological or logistical system, but transcend the boundaries of various systems, and are also a social system. Hence social systems, later called industrial symbiotic networks, may be considered as social networks. This implies that links in industrial symbiotic networks in firms, households, policy actors, and social actors operate in the context of bounded rationality, i.e. they are limited in their rational decision-making not only by their own inherent organizational values but also constrained by the decisions of other organizations (Džajić and Rončević 2017). Since exchange, the elementary activity in industrial symbiosis, is taking place between firms and social actors and it requires cooperation, communication, and a certain level of trust its morphogenesis can be studied. Forming an industrial symbiotic network is a complex process, and its success is highly dependent on establishing cooperation across several spheres including physical, social, and organizational ones. Neglecting even one of them may lead to a significant time delay or an even frailer development of a useful industrial symbiotic network (Džajić and Rončević 2017; Boshkoska, Džajić and Rončević 2015).

3.1 Morphogenesis of industrial symbiotic networks

We discover that symbiotic cooperation between participating firms in respective case studies is like »Morphogenesis«, from the Greek: *morphê* as shape and *genesis* as a creation, literally, »beginning of the shape« (Bard 1990). The morphology of social networks could be an aspect of how industrial symbiotic networks are composed of systems and sub-systems that operate interdependently.

> *The stages of paradigms and the state of hypothesis are all based on the view that organismic development is a useful analogy for the growth of firms inside industrial symbiosis. Often, this analogy is taken directly from the human experience of aging: »The life-cycle approach posits that just as humans pass through similar stages of physiological and psychological development from infancy to adulthood, so businesses evolve in predictable ways and encounter similar problems in their growth« (Bhidé, 2003, 244). Overall, the core assumption in this paradigm is that: organizations interacts if they are developing organisms* (Tsoukas 1991, 575).

Industrial symbiotic networks are complex systems constructed by humans; they represent culture, ideas together with the flow of people, and they regulate network growth and its development. It is hence critical to understand patterns: shapes of flow, and how the difference of networks and the morphology can adapt or transform these flows. Morphogenesis is the process through which an industrial symbiotic network comes into existence as an object and develops its patterns. Building morphology as a collective social system will be developed as a model through this study, capable to implement such patterns (Džajić and Rončević 2017).

Interactions are compared to morphogenesis, like a biological process that causes an organism to develop its shape. The theory here is that the morphogenesis can take place also in a social system of industrial symbiosis in social culture and/or inside masses of industrial symbiotic networks. Morphogenesis, in this study, also defines the development of interaction forms, which can describe the evolution of a structure of social aspects of industrial symbiosis (Müller 2013, 60).

3.2 Mutual environmental society: Clusters theory

Corporate environmental management literature presents many terms and concepts to clarify how collaborative actions of firms on environmental issues could contribute to natural resource preservations. Some so-called terms are eco-clusters, industrial ecology, industrial symbiosis, corporate environmental network, and eco-parks. Although the focus of the formation of those types of firms concerns solving environmental problems, nevertheless they provide essential conditions for creating innovation and improving competitive advantages. Esty and Porter (1998, 35–43) stated that »industrial ecology thinking would assist firms to make productivity improvements in using raw materials and therefore improve their competitiveness« (Doménech 2010, 152).

In industrial symbiosis, a feature of this production system consists in many small- and medium-sized firms, which are linked to the production of the same product but on different production levels, with different tasks. Industrial districts consist of between 500 and 3,000 firms with 10,000 to 20,000 employees, who come from relatively small local communities and usually specialize in the production of a single product (Ješovnik and Jaklič 1998, 57–58).

Industrial symbiosis can also be defined as a different form of links named as such: circular entrepreneurship, the creation of a joint venture, consortium, dynamic network, industrial cluster, industrial district, or rayon (industrial districts are created because of the unique industrialization, therefore spatially limited local community, which is a common manufacturing system).

The definition of the system as a cluster is the closest in meaning to the system of industrial symbiosis. The cluster theory is a group of organizations and institutions related to a geographical environment that brings together manufacturers, service support, suppliers, educational institutions, and commercial organizations.

Clusters derive from alliances or cooperation and the combination of entities involved in a given area. This group of entities hold a common interest. Porter defined a cluster as a geographical concentration of informally associated organizations, specialized suppliers, service providers, organizations of similar activities, and institutions in a particular field within the framework that can lead to cooperation and competition. Porter, like many other authors, also stressed the importance of the location (1998, 179) and leaders willing to participate.

Synergies start partly because most of the higher management participate in social activities (»social clubs«). This is felt particularly in the independent private sector, which is almost entirely focused on market forces (Boshkoska, Džajić and Rončević 2015).

Economic development nowadays opens new ways of doing business, which in the past were not possible or only with high transaction costs. The new economy brings new opportunities for producers of industrial policy. Support for the introduction and establishment of network connections, clusters, bring a new kind of politics, represented by numerous designers of industrial policy. In many countries, in different industries, this type of integration has already shown good results. Firms and the environment are not always prepared to change and adopt integration processes. The reasons are ignorance, unwillingness of firm owners, and firms in ignorance of options or the adequacy of the environmental conditions or the incompetence of the policy makers.

Theoretical concerns about the meaningfulness of those of network clusters can be seen in the theory of Weber and Friedrich (1929), which speaks of the dependence between the transport and international trade, so that the industrial body works better near the main markets and centres. Weber and Friedrich emphasizes the transition from traditional (rural) to industrial (urban) society that defines the social and spatial organizational streamlining and trade (market).

Rationalization is expressed through the mechanisms of effective regulation and administration (bureaucracy). The theory is supplemented with the concept of increasing the internal and external benefits and imperfect competitive advantage brought about by geographical and historical factors.

Therefore, industrial environments such as industrial symbiosis are the spatial expressions of rationality. It is worth mentioning that one of the representatives of contemporary sociology, Anthony Giddens (in Giddens 1989), presents the theory that the intersection of the social structure and the way operations are structured is the starting point for studying the socio-spatial relations. Industrial spatiality should be studied as the interplay between micro- and macro-sociological phenomena in space and time. The process of temporal and spatial distinction that makes relative physical proximity, presence/absence, and timing characterizes modern society.

3.3 The drawback of sustainable embeddedness in industrial symbiosis

Researchers of regional socioeconomic systems have hypothesized that the transition from non-ecological (called unsustainable process in this context) to ecological (called sustainable process) is an evolutionary process, most likely to be introduced at the local or regional level (Wallner and Narodoslawsky 1996, 221–240; Wallner, Narodoslawsky and Moser 1996, 1763–1778).

Sustainable development requires the selection of an optimal system boundary by actors when they develop goals and form clusters (Baas and Huisingh 2008, 399–421). This requires the ability to look at activities in terms of the system in which they are situated and evaluate them accordingly. Given the selection of a system boundary, actors need to be able to mobilize the players that are part of the present or envisioned future system. This can be called the actualization of the selected system. In a regional cluster, there will be actors that do not wish to cooperate because sustainability is not their aim. The number of firms, their diversity in size and type, and the intensity of their interactions are major variables in the system. The different dimensions of firms can be described in terms of cognitive, structural, cultural, political, spatial, and temporal established-embedded (firmly and deeply rooted) (Baas and Boons 2006, 1073–1085). Simsek et al. (2003) use in addition to relational embeddedness of firms, as they are described regarding the quality of the relationship highlighting the effects of cohesive ties amongst social actors on their economic activities. They link relational embeddedness strongly to structural (Granovetter 1992, 3–11; Gnyawali and Madhavan 2001, 431–445) and cognitive embeddedness. Uzzi (1997), which in this study on embeddedness, creates the following paradox. The stronger the embeddedness, the more difficult it will be for the counterparts to change to other partners, at least in the short term. The weaker the embeddedness, the more the relationship will have an arm's-length characteristic – shows one of the illustration based upon 12 years of experiences in the Rotterdam Harbour and Industry Complex (Baas and Huisingh 2008).

3.4 Exchange conditions and social mechanisms of control

There are two key ingredients for the success of industrial symbiosis (Gibbs and Deutz 2005; Tudor, Adam and Bates 2007): firstly, the exchange of material by-products and energy; secondly, some form of social exchange, in other words, inter-firm networking, trust, and collaboration. In this section, we will take a closer look at the latter component: What role do social exchanges play in networks that are already engaged in physical exchanges?

Yes, there is the importance of the social dimension of the industrial symbiosis acknowledged in literature, but there is still a lack of a certain level of the research and study of it. Trust and willingness to cooperate are seen as requirements for »industrial symbiotic appropriateness« (exchange of materials and energy) to take place (Ehrenfeld and Gertler 1997; Gibbs 2003, 222–236; Gibbs and Deutz 2005, 452–464; 2007; Hewes and Lyons 2008; Tudor, Adam and Bates 2007). Many failed attempts exemplify the importance of these factors to intentionally plan and design eco-industrial parks from scratch and through policy intervention (Chertow, 2007; Gibbs 2003; Gibbs and Deutz 2005, 452–464, 2007, 1683–1695). The reason many of these industrial symbiotic projects fail is that it is hard to find firms willing to co-locate and link their processes with other firms they do not yet know or *trust*. This is perceived as simply too great a risk to take. Trust is needed before interdependencies though by-product exchange can be set up, because certainty and continuity of supply are extremely important to industrial firms (Tudor, Adam and Bates 2007, 199–207). Also, trust is needed in exploring the possibilities for by-products exchange, because firms need to share (possibly sensitive) information about their inputs and outputs (Ehrenfeld and Gertler 1997, 67–79).

The social aspects of industrial symbiosis is still very much »under theorized« (Doménech and Davies 2011, 79–89; Gibbs 2003, 222–236; van Koppen and Mol 2002, 132–158). The most recent and comprehensive attempt to address this knowledge gap is an article by Doménech and Davies (2011), in which they try to explain the emergence and evolution of industrial symbiosis in pre-existing industrial symbiotic networks, including the social mechanisms involved.

Central to Doménech and Davies' analysis (2011), for example, is the concept of »embeddedness«, which refers to »the mechanisms through which social structures, cognitive processes, institutional arrangements and cultural context determine the action of economic and social representatives« (2011). These embedded networks have four main characteristics: fine-grained information transfer, multiplicative of personal relations, joint problem solving, and trust. These characteristics in the embedded industrial symbiotic networks enable the emergence of new capabilities that are not specific for individual firms, but are shared by the network. Baas and Boons »allow firms to be more flexible and adapt more quickly in environments characterized by complexity and continuous change« (Baas and Boons 2004, 1073–1085). (Doménech and Davies 2011, 79–89). Doménech and Davies identify three phases of cooperation through which industrial symbiotic networks have become increasingly more embedded: emergence, probation, and development and expansion. In the progression of these phases, the four characteristics that facilitate this embeddedness – and thus the build-up of shared dynamic capabilities – also become increasingly manifest.

In the emergence phase, the first relations between actors are developed and some straightforward opportunities for collaboration are explored. Generally, these first steps do not require large structural changes in processes or technology, but they do form a basis for trust and further cooperation. What follows is the probation phase. After some time, actors get a general idea of the dynamics of the network and of the opportunities for exchanges. This will lead to more exploratory collaborative projects. Once there is enough trust to experiment and more successful linkages are achieved, this in turn creates more trust: »it is the positive experience of interacting over time that achieves trust between actors and leads to the potential for more complex embedded networks over time« (Hewes and Lyons 2008, 1329–1342). The test phase takes time and can gradually result in more integrated, shared decision-making routines between actors. If the second phase is successful, what could potentially follow is a phase of development and expansion. »Continuous collaboration, communication and increase of experiences of firms allow the increasing of embedded connections (ties), governed by trust, unspoken knowledge, joint problem-solving, and generate routines of cooperation that meaningfully reduce the transaction costs related with it« (Doménech and Davies 2011, 304).

Such a policy actor can have an important role in strengthening network ties and deepening the embeddedness of the network (Chertow 2007; Doménech and Davies 2011; Ehrenfeld and Gertler 1997, 67–79; Gibbs 2003, 222–236). Concepts that bear resemblance to our approach include knowledge »spill-overs« and coopetition. Coopetition can be a hybrid between competition and cooperation amongst firms, which results in a complex relationship with contradictory logics of interaction. Additional examples of these are also strategic alliances between multinational firms or multi-unit organizations where different brands opting for the same market belong to a mother firm (Bengtsson and Kock 2000, 411–430; Kilduf and Tsai 2003). Knowledge »spill-overs« refer to the existence or emergence of shared competences between firms in the same industry, located within the same industrial cluster or district. These shared competences consist of knowledge flows (both explicit and implicit) concerning products, processes, technologies, consumers, and markets that are accessible only to firms within a district (Camisón and Forés 2011, 66–86). They are intangible assets that are hard for outsider firms to imitate, appropriate, or substitute because they are »largely district-specific, idiosyncratic and based on tacit knowledge, unique institutions and multiple links between actors« (Camisón and Forés 2011, 66–86). Knowledge »spill-overs« emerge because of multiple factors: local availability of highly qualified, specialized, and experienced human capital; local mobility of labour; local spin-off firms; informal social links between local employees of intra-district firms; and the existence of a community with commonly agreed upon standards of behaviour, embedded in local traditions (Camisón and Forés 2011, 66–86).

Due to the competitive element, the playing field to evolve relationships and develop trust amongst actors is limited, because competing firms take a thoughtful position. This is a clear difference with *networks*, which are usually characterized by the absence of competitive threat due to the different industries in which firms are operating. However, when firms in similar industries are part of a larger network, interactions between those firms could take the form of coopetition. A requirement of effective coopetition is either a clear internal separation of competition and coordination activities, or the simplification of cooperative aspect by an external actor such as a collective association (latter being common within industrial symbiotic networks) (Bengtsson and Kock 2000, 411–426).

Interconnections with embedded networks are the existence of recognized knowledge transfer and the multiplicity of relations, but other aspects crucial to the emergence of shared dynamic capabilities in these networks, such as trust and joint problem solving, are not at all emphasized in the literature.

3.5 Building a complex system approach

The environmental, social, and economic development together forms a sustainable development. Economic development is the cornerstone of wealth creation, but this is subordinate to and dependent on society and the environment. The existence of industrial symbiosis therefore consists of the individual actors bringing mutual benefit, with an emphasis on reducing the consumption of energy, materials, and proximity/vicinity and increasing the dependence of internal participants in firm. The occurrence of industrial symbiosis usually derives from the local or regional environment because the industrial environment offers natural capital and affects the development of social and economic perspective.

Cooperation between firms creates a large circuit that control costs, and greenhouse gas emissions are reduced in the process. This symbiotic cooperation between firms also has an impact on reducing the consumption of raw materials and energy, as these are replaced only with products that are incurred in the territory of the participating industries. Positive social impacts are reflected in the increase of the number of jobs requiring a higher level of education. Industrial symbiosis creates prosperity of local society as it primarily promotes cooperation between small businesses with small local providers.

The business environment greatly influences the possibilities that this environment offers. In addition, firms and other organizations also try to make a more effective use of this environment. The tradition here is also linked to industrial, innovation, and local politics in a given environment. Development and forms of cohabitation affect environmental standards, environmental stability, and regulation, taking into account the autonomy of local authorities – the importance of the attitude of the state of integration policy. Features such as a positive attitude and support of complex forms of links between the parties, technology, raw material suppliers, the pooling of competition, and the desire to progress by local and central government using the industrial symbiosis greatly facilitate the creation and development of this symbiosis. Tradition, therefore, is not unaffected by suppliers, raw materials, access to technologies, and tolerance to changes in the environment. This shows as in recent decades many attempts at the implementation of such schemes, of which not all have achieved self-sufficiency situation (automotive industry, biogas, etc.,) have been witnessed.

Indeed, the complex rate of such organization does not give priority only to physical obstacles but to social ones as well, such as commercial confidentiality issues, the lack of mutual trust between firms, or a lack of strict environmental regulations. It is true that without the support of the municipality and the construction of such models' industrial symbiosis, it would be difficult to implement. As an important mechanism to increase the efficiency of the cooperation awareness, synergies should take place already at the superior (CEOs) level of firms. At this point, there should be an awareness of »disclosure«. Logistical long distances between undertakings may affect decision-making, while the business within a short distance can facilitate the feasibility of the development process.

The true meaning of industrial symbiosis, and in some cases the practice, is losing the meaning because it is dominated by economic rather than environmental aspects. The process between producers/suppliers is a set of agreements formed by the usual manufacturer in the relationship with the supplier. When the cooperation developed between the two, a gap in terms of communication or in another organizational perspective appears; thus, creating a common communication and organization platform.

These common communication platforms are treated as the interaction between people, representatives of firms (leaders), but also as people who are setting up mutual interests, trust, and liabilities as a key mechanism for social interaction. In the case of confidential information and competition, this form of social relationships can become negative and risky. The bilateral agreements and social interactions, the collection of trust, good communication, the common platform system of aid, the issue of confidentiality, and the associated risk should all be taken into consideration by participating organizations who undertake symbiosis (Almasi et al. 2011).

As an exercise of common interests, cooperation and networking in industrial symbiosis, businesses play a special role with respect to:

- The potential for environmental improvement, that is seen to lie largely with technological innovation;
- Businesses as a focus of technological expertise as an important representative for accomplishing environmental goals;

- Command-and-control regulation as being greatly inefficient and, at times, counter-productive. »Perhaps more significantly and in keeping with the systems focus of the field, many see industrial ecology/industrial symbiosis as a means to escape from the reductionist basis of historic command-and-control schemes« (quoted in Lifset and Graedel 2002, 8; Ehrenfeld 2000, 229–244).

Due to communication platforms, those symbiotic actions (such as reducing inputs, production, and waste management costs, and by generating additional income due to value added to by-product streams, reducing resource use, improving relationships with external parties, and by facilitating the development of new products and their markets) generate new employment and help to create a safer and cleaner natural and working environment. The well-being of a growing number of people depends on resources that are diminishing. At the same time, production chains and trade have become global, as have the environmental effects of production and consumption, such as climate change (Chertow 2007, 11–30; Solomon 2007). Chertow (2007) also sets up three main common interests when giving incentive for resource sharing:

- The integration of local measures in the national strategy should not be a return to the centralized management of local development policies and decreased autonomy places;
- Principles which lead to different actors in the implementing of applications and synergistic activities and thereby improving ability to achieve common goals, facilities/infrastructures sharing and;
- The improvement of environmental conditions reinforces innovation and the ability of firms and their international influence in the region/country.

For example, Korhonen and Snäkin (2005, 169–186) analyse and explain the evolution of networking in industrial parks using diversity, and connectance. They argue that increased diversity (of the actors involved) enhances connectance and opens new possibilities for cooperation, although increasing the number of actors can also lead to conflicting interests, thereby preventing connectance and interdependency.

As Spekkink (2015, 29) in his opinion found, that Ashton, in her methodology builds the analysis of the structure, function and evolution of regional industrial ecosystems by combining perspectives from economic geography and industrial ecology with Holling's theoretical structure of complex systems. The prominence in Ashton's structure is on successional changes observable in regional industrial ecosystems (exploitation, conservation, release, and mobilization). Patterns of change in regional industrial ecosystems are described for multiple levels of analysis, using the comparison of ecosystem development and concepts borrowed from literature on economic geography and industrial ecology (Spekkink 2015, 29).

Mirata and Emtairah (2005, 993–1002) discuss industrial symbiotic networks from the perspective of innovation studies. The authors argue that industrial symbiotic networks can contribute to fostering environmental innovation at the local or regional level by stimulating the collective definition of problems, providing inter-sectorial interfaces, and promoting a culture of inter-organizational collaboration oriented towards environmental challenges.

There are several publications that draw more attention to the social aspects of industrial symbiosis; Baas and Boons (2004, 1073–1085) develop a social science structure for investigating regional industrial ecology. »Posch investigates whether industrial recycling networks or industrial symbiotic projects can be used as a starting point for broader inter-firm cooperation for sustainable development« (2010, quoted in Spekkink 2016, 29). Lambert and Boons (2002, 471–484) describe the sustainable development (including industrial symbiosis) of industrial parks as a social process based on ecological, social, and economic aspects and emphasize the importance of learning processes among social actors.

Ashton (2008, 34–51) discusses the increasing attention given to the social aspects of industrial symbiosis and introduces the analysis in the field of social networking as a methodology to study the patterns of relationships, interactions, and social structure that are often emphasized in social theories. She demonstrates the relevance of this methodology in the context of industrial symbiosis by using it to analyse the relation between connectivity amongst firms and managers. Wright et al. (2009, 551–564) propose a methodology for translating ecological quantitative analysis techniques to an industrial context. The methodology is demonstrated in one of the cases of industrial parks in Canadian urban neighbourhood, located along the northeast shore of Bedford Basin of the Halifax Regional Municipality in Dartmouth, Nova Scotia. They were using the concepts of connectance and diversity. The demonstrated techniques can potentially aid in gaining an understanding of industrial symbiotic networks.

In this study, conceptual structure connects well to the existing conceptual and empirical literature on industrial symbiosis and their organizational and social aspects. It provides a way to further develop the field by building on theoretical insights that help to understand. In the dynamics through which regional industrial systems change their connectedness consequently affect their ecological and sociological impact (quoted in Ayres 2002, 114–171; 2002).

3.6 Building trust, good communication, and proximity

Having, for example, a bilateral agreement surrounding a cooperation implies the need to know each other's quantity and quality of material required, technology used, production processes, waste, etc. Social interactions gather effects such as *trust, good communication,* and *proximity* required to cover confidentiality issue–related risks (considering the specific nature of waste being exchanged: communication and trust are thought to be important when the materials being exchanged have potential liabilities because they are subject to environmental regulations (Chertow 2004)).

Kurup (2007, 19) states that: »social benefits of industrial symbiosis materialize through trust and cooperation between partners or organizations taking part in the symbiosis«.

Trust between firms decreases operation costs, risk, and the uncertainty of the exchange, while being extremely important for the development of cooperation structures and cooperation itself (Gibbs 2003, 222–236; Chertow 2007, 11–30; Baas 2008, 330–340; Ehrenfeld and Gertler 1997, 67–79 quoted in Doménech 2010). A lack of trust within industrial symbiosis complicates and inhibits communication and motivation for cooperation between economic actors (Gibbs 2003, 222–236; Chertow 2007, 11–30; Doménech 2010). Accordingly, this study represents the basis for finding causality between social aspects, whereas we need to understand that there is no universal causality and that no causality can be applied to all instances, neither in industrial symbiosis nor in networking between irrational societies. When both communication and motivation exist between industrial symbiotic firms, trust and cooperation can be seen as the main social aspects for the realization of its ecologic and economic goals. Communication and motivation as part of industrial symbiosis can be viewed as social aspects, as factors of influence on social aspects, or even as the conditions for social aspects (Doménech 2010). In our research, we have focused on two social aspects, trust and cooperation, representing key factors of the successful and effective exchange in industrial symbiosis. With regards to trust, we have focused on inter-organizational trust, because industrial symbiosis not only implies trust within one organization, i.e. organizational trust, but demands trust between several organizations or firms. Despite focusing on inter-organizational trust, we accept that interpersonal trust within one organization can result in organizational trust and, consequently, in inter-organizational trust. Considering cooperation, we focus on inter-organizational cooperation, applying the same principles as for trust, because industrial symbiosis not only involves cooperation within one organization unit, but also wider, i.e. between several organizations or firms. »Trust represents a method of establishing a relationship and a means for its preservation. Trust is a simplification strategy, allowing us to adapt to the complexity of the social environment we live and work in« (Sztompka 1999, 25).

The processes of entrepreneurship result in the emergence of new products, services, and value creating systems particularly in social contexts, for example, imposing certain roles that should be played among firms. An important task of firms is to adapt the organizational structure of any changes in the organization and the environment in which it operates. Another important task is the adaptation of the organizational structure. Firms must achieve consistency and adapt the organization. Of course, they must be aware of limitations imposed by the employees – even in the freedom of differently organized structure of firms.

Firms are not independent units, isolated from the rest; their importance is clear and their interaction with the environment must be constantly maintained. Thus, industrial symbiosis represents a system, an organized body, which consists of parts or more mutually dependent parts, components-subsystems, and is separated from its environment by the boundaries, which can be easily determined.

While the dimension of actors' entrepreneurship suggests that acts of creativity are significant in initiating change, the emphasis on context too is very significant. To remain fit over time in the dynamic systems, it is essential that creative firms constantly organize novelty in anticipation of new collaborations, new networks, and new patterns of consumer behaviour. An obvious approach to dealing with this fluidity is to simplify study designs by focusing on one level of analysis, in most cases either the individual, the firm, or the industry.

Theoretically, it is considered that such temporary structures have ontological position, as ideas, conceptual models, thought experiments, spontaneous connections, and fleeting alliances attract and amplify changes of creativity, energizing, new patterns and new possibilities.

Although economical profitability is a strong incentive pushing towards cooperation, it does not cover risks relating to confidentiality issue (van Beers et al. 2007a, 830–841).

Firstly, we must mention the social capital theories, developed by different authors such as Granovetter (1997, 1977, 347–367), Bourdieu (1986), Putnam (1994), and Coleman (1994), while differing widely, all tend to highlight (in the framework of an inversion of power relationships among actors): the importance of trust-based interpersonal relationships and social networks, in promoting social solidity, fostering economic development, and how to manage crisis situations (Ricerca–azione sulla povertà e l'esclusione sociale 2015; Andrea, Quaranta and Quinti, 2005; Archer 1995, 2008, Pact-carbon-transition 2010; Rončević and Makarovič 2010, 2011).

Social interactions occurring between the actors of an industrial symbiotic system leads to a necessary cohesion to develop further willingness to invest into industrial symbiotic synergies. When such situation industrial symbiosis is reached, the following interactions between actors and structures are: interactions between operating firms in the area and social interactions between firms, between leader staff to maintain trust and willingness communication, and when processing operations involving sensitive business information and environmental risks (Džajić and Rončević 2017).

The attractiveness of the environment is the result of a »clear« circle: the environment provides a range of economic factors (tangible and intangible) for actors, that are important for the development of resources and skills needed to achieve a competitive advantage over competitors located in other places. They all have large and long-term impact on the local economic and social dynamics at the local intellectual capital and in some cases on the landscape.

Going deeper into the research, we can only mention and someway compare connections in the industrial symbiotic networks as links between structures and agencies as basic concepts of sociology. Firstly, the concept refers to bounds of social nature (behavioural patterns, norms, social rules, organizational procedures, dominant social representations, etc.,) which channel and condition individuals (both in cognitive and operational terms), while the second concerns the orientation of individuals and organizations towards action, which is manifested in intentionality, plans, lifestyles, or forms of social mobilization. Of key importance is the evolution of the relations between structure and agency, examined by Quaranta (1985), Giddens (1994), and Archer (1995, 2008), especially the transition from »contemporary society« to the so-called postmodern society.

> At the heart of this transition there is a gradual difficulty in the relative strengths of structure and agency, producing a progressive weakening of social structures and a parallel overall strengthening of agency. The deep changes are instantaneously affecting all economic, political, scientific, ethical and cultural institutions of contemporary society; and the increasing amounts of independence people have in choices and actions, either as individuals or in organizations« (quoted in Pact-carbon-transition 2010, 21).

The results of this problem may be seen in such phenomena as the increased uncertainty that characterizes the lives of individuals, the increasing predominance of horizontal relational networks over hierarchical relationships, the fragmentation of lifestyles, the increased social segmentation, or the acceleration of all social dynamics (institutional, technological, economic, or organizational) (Archer 1995, 2008).

The activities and production are goal-oriented, as they are directed at finding congruence among parameters synthesizing coherent functions, which the goods should perform. In performing them, which is in searching for solutions, however, an economic involver might, but usually do not, have all the relevant information and thus need to act adaptively by interacting with other firms. An important qualification in the way firms interact in the production process emerges by recognizing that search activities are rarely completely random and undirected, since adaptive actors create and employ cognitive mechanisms for ordering the world. On the contrary, drawing on the evolutionary theory of human cognitive processes (Cosmides and Tooby 1992 quoted in Barkow, Cosmides, and Tooby 1995), we maintain that individual organisms (and their minds) are aggregates of information-processing mechanisms, chosen by natural selection through adaptation, search, and discovery of »statistical regularities«, in an environment which would otherwise be a source of insolvable computational problems. The economic space that we examine is composed of interactions between adaptive actors who operate within an abstract sequence of phases/operations, the output of which is a vector of attributes. This identifies for them an action space, a »connective geometry« (Potts, 2000, 2001, 413–431). Indeed, in this action space, the geometry of connections determines a field of inquiry for firms, with respect to the different situations that can be envisaged. In general, however, everyone interacts with everyone else, so that an exponential number of alternative combinations can be conceived. When repeated feedback and cycles of interactions occur, networks emerge and entail a complex kind of dynamic (Leoncini, Lombardi, and Montresor 2009).

»Relationships between the firms constitute the »weave« within which their heuristics are formed, sifted, sorted out, shared, reconbined and, possibly rejected. The connections among adaptive entities give rise to different relational forms that is to different evolving morphologies« (Cosmides and Tooby 1992, 163–228). If we assume that interactions in industrial symbiotic networks play a *very* important role in contemporary capitalism, then the critique of these phenomena in contemporary society becomes one of the tasks of a critical theory of society.

By this, the social structure in industrial symbiotic networks is meant as a set of rules, analogous to the paradigmatic structure of language, the virtual structure of significance« which provides the underlying grid from which actual speech is generated. The concept of »activity dependence« is poorly defined and potentially misleading. Archer leaves us free to observe empirical relationships in a realist manner without committing ourselves to what see Bhaskar in his metaphysics. Giddens gives us a theoretical vocabulary that tries to capture the relationship between social systems and the involved actors

*who make them up. What we see are »syntagmatic instances of the paradigmatic struc-
ture. For Giddens, social systems correspond to this syntagmatic dimension. They are
actual patterns of interaction and observable social relationships* (Giddens 1979, 15
quoted in Healy 1998, 510).

*Some aspects of social structure seem irreducible to the conceptions, beliefs, or deliberate
actions of actors. On the other, no one wants social structures wandering around by themselves like so many lost cows. The solution seems to be to find a way of understanding structure that avoids making it reducible to rules* (Giddens 1979; quoted in Healy 1998, 515).
*In the language of critical realism, it shows that the solution seems to be to come clean and
argue that such systems are not virtual at all, but actually exist. Societies in industrial symbiosis are emergent from individuals and real in themselves: they refer to radically different
things* (Bhaskar 1989, 33 quoted in Healy 1998, 515–516).

*Archer's determination to develop a comprehensible and empirically gainful view of struc-
ture and agency has much to recommend. Her alternative has several advantages over that
of Giddens, not least a strong effort to link theory and research. The approach proceeds on
two fronts. The first, »analytic dualism«, is the claim that effective sociological research
depends on a clear distinction between actors and structures also in industrial symbiotic
networks* (Archer 1995, 158 quoted in Healy 1998, 516).

The second front, Archer's claim is that:

*our explanations will be unable to do justice to what we observe unless, for the sake of
analysis, we think of societies and individuals as different things* (Archer 1995, 158,
quoted in Healy 1998, 516). *But, in addition to this, Archer also »promotes a specific
ontology, critical«* (Bhaskar 2010, 2014 quoted in Healy 1998, 516). *The two fronts are
then merged into a paradigm which states that the »analytic dualism relies on this brand
of realism«* (Healy 1998, 516).
*The parts of Archer's account of structure does not clearly explain how individuals and
societies are related to one another. Her concept of »activity dependence« is especially prob-
lematic and what Archer has had in mind all along. Archer needs structures to be real,
and so binds her arguments to Bhaskar's ontology (Archer 1995, 135–162). We can move
towards a useful, non-productive thought of physicalism of societies as abstract objects
made up of relations that supervene on individuals. This allows that social structures can
have casual properties of their own, but makes it very clear what their relations to individual actors must be like (Healy 1998, 516).*

This industrial symbiotic network chain is very complex, but there is no conceptual mystery about it. However, such causal explanations form no part of the
conceptual relationship between actors and structures. As it exists today, the
demographic structure supervenes on those individuals. Archer believes that
the concept of structure should not be sacrificed to that of agency. This »analytical dualism« is absolutely necessary for the human actors' creative capacity
to distance themselves cognitively from the objective social and cultural
circumstances, to critically see and elaborate on them (meta-reflexivity). In other

words, it is necessary for the sequential radical conversion of the »individual actors« into a self-effacing social actor and role-taker (Archer 2003).

3.7 Jens Beckert's theory: Social forces develop networks in industrial symbiosis

The development of industrial symbiotic networks research in this present study is based on Beckert's social fields theory (Beckert 2009, 245–269), according to which social fields, which in our case are industrial symbiosis and its networks, are shaped by the three social forces. Beckert (2010) acknowledged three types of social forces: *social networks, institutions*, and *cognitive frames*. Institutions as laws, central concern for law, the formal mechanism for political rule-making, and enforcement; social networks as social structure made up of a set of social actors such as individuals or firms, a set of the dyadic ties between these actors; and cognitive frames as social interaction, meaning-making technologies, and strategically selective opportunities for reflection and learning (Beckert 2009, 245–269). This being relevant for explaining also economic and technological outcomes of this study model. In his social framework background, he explained that »these forces have been applied to diverse economic facts (as the level of effectiveness of economies, the stratification of opportunities in labour markets or the formation of prices (cf. Beckert 2010), although occasionally considering their influence on each other« (Beckert 2010, 606; Fourcade 2007, 1015–1034). »Those interrelations among social forces effect market dynamics, since they are responsible for the position of actors in more or less powerful positions as well as for their resource endowments. Those are important preconditions for actors to influence the social forces in accordance with their interests« (Beckert 2010, 611).

Besides social forces, the environmental and organizational economic sphere provides a comprehensive tableau of the social composition in a common market field of industrial symbiotic networks. Other factors that influence market outcomes independent from the social composition of the field are technology and resource scarcity. The analysis of technology has become a prominent research object in the social studies (Preda 2007, 506–533); the effects of resource insufficiency remain typically external to the scope of market sociology.

Actors inside industrial symbiosis gain resources from their position which they can use to influence institutions, network structures, and cognitive frames.

To simultaneously consider all three social forces in market fields and their reciprocal influences allows us to consider their interrelations as sources of field dynamics. While it might be useful to distinguish the three structural forces analytically, any approach that does not consider all the forces influencing action remains necessarily incomplete in its analysis and is in danger of drawing a distorted picture of the embeddedness of economic action and the dynamics of market fields« (Beckert 2010, 665).

However, it seems too deterministic and too static. It is too deterministic in the sense that it suggests, despite its focus on »ongoing systems of social relations« (Granovetter 1985, 487) that we grasp economic outcomes simply as a function of social structures without bringing into focus the agency processes through which actors interpret social structures and which lead to contingent responses to the situation. It is too static because it does not indicate the mechanisms through which the embeddedness of economic action changes (Džajić and Rončević 2017).

Based on the field concept, Beckert discusses the interrelationships between the three types of structures identified and their role in the change of market fields: »markets as institutions« (Barmer, Naazneen and Vogel 2007; Fligstein 2001), »markets as networks« (McLoughlin 2002), and »markets as cultures« (Abolafia 1998). Sociological approaches, basically towards the economy, explain economic outcomes based on the influence of social structures on individual action. By explaining the resolution of coordination problems and distributional outcomes by means of the social forces entangling market actors, sociology provides an alternative to those economic approaches that proceed based on the individual interests of actors when addressing the question of economic order.

»While experts focus on the structure of social relations and their networks (Burt 1992; Granovetter 1985; Uzzi 1997; White 1981), they often consider the role of institutions and cognition in their explanations of economic results« (Beckert 2009, 247).

»Networks in industrial symbiosis are mostly based on the idea that the most important components of social life rest neither in the formal institutions under which actors operate nor in the individual attributes and traits with which they are identified« (Mizruchi 2007, 7). Cognitive frames have been brought into institutional theory and analysis by combining them with networks and institutions. This can be seen in sociological institutionalism, which emphasizes the role of cognitive frames, meaning structures as decisive for the explanation of economic outcomes by broadening the conception of institution.

Different types of social forces fail to take into consideration their analytic disconnection (Archer 2003). The irreducibility of social structures has been acknowledged in a more comprehensive sense by inquiries which systematically investigate the influences of other types of social structures on the development of the structure in focus. Beckert contributed with a systematic tableau of these influences, based on the premise of the irreducibility of different social structures, instead of conceptualizing an isolated influence of one social structure on one other. The concentration on a single macrostructure gives support to deterministic understandings of this structure. The analysis follows a »morphological determinism« (Fourcade 2007, 1021) through which action is subsumed under social structure. »The actors, their dimensions, and what they are and do, all depend on the morphology of the relations in which they are involved« (Callon 1998, 8).

To conceptualize rationally, the common influences of the three social structures and the relationship of structure and agency needs, however, an overarching conceptual framework that can encompass the different types of structures and can allow the conceptualization of their interrelations (DiMaggio and Powell 1983, 147–160).

»The multiplicity of social forces structuring a market is a source of stability if the different structures reinforce each other. But it is also a source of instability in markets if changes in any one of the structures have repercussions on the strategies of actors because of newly emerging profit opportunities and the closure of hitherto existing ones« (McDermott 2007, 885–908 quoted in Beckert 2010, 614). Though cognitive frames are in many ways »given« and escape manipulation because of being taken for granted, institutional entrepreneurs can pursue strategies that aim at what could be called »cognitive hegemony«. »One central activity of institutional entrepreneurs is to provide and secure the ideological grounds on which institutional regulations advocated by them find legitimation« (Beckert 2010, 616).

Beckert explains in his theory that:

> cognitive frameworks with social networks are another structure that impacts the reproduction and change of institutions. Cognitive frameworks exercise their influence through the ways in which they constitute the perception and the legitimation of institutional forms and network structures. They are part of market fields and must be investigated in their interaction with networks and institutions. Institutional entrepreneurs are embedded in these social structures. Taking this into consideration makes it also possible to move away from individualistic concepts of agency in institutional theory that square oddly with the basic premise of the social embeddedness of economic action (Beckert 2010, 617).

»The stability and change of cognitive frames, however, can also be understood within the two other social structures discussed, i.e. institutions and social networks« (Beckert 2010, 218). In more recent institutional scholarship, the role of socialization has been discussed using the notion of »normative pressures« (DiMaggio and Powell 1983) exercised by the socialization of actors on the

norms of their profession, which is one mechanism leading to the stabilization
and homogenization of fields.

> *The influence of institutions on cognitive frameworks takes place through socializing
> institutions like universities, business schools and professional associations which shape the
> cognitive frames of them. This can either be a stabilizing influence on dominant frames or
> one that transforms prevalent cognitive frames if the actors entering the market field have
> been socialized into a different mind-set than the executives* (Déry, Mailhot and Schaeffer
> 2007 quoted in Beckert 2010, 618).

The firms must overcome different social barriers. It has been found that the
physical dimension of such a system in industrial symbiosis, relating to the technical feasibility of synergies, only comes as an achievement of previously accom-
plished interaction and social work. Societies will be bound to have collective
symbiosis that is, at least, the sum of the individual metabolisms of its actors. If
a society cannot maintain this metabolic turnover, its population will, unfortu-
nately said, emigrate.

This interdisciplinary study is becoming increasingly relevant since the inten-
tional or planned development of industrial symbiotic networks has immense
potential due to their contribution to the reduction of entropy via savings in
energy and materials, and reduced waste production, hence its contribution to increase in value-added and quality of life. These results are achieved
by improving the environmental performance of industry, by the increasing
effectiveness of technologies and processes for the reduction of environmental
damage, and especially by connecting flows of materials and energy, within the
framework of industrial symbiotic networks.

Industrial symbiotic networks as a collective system of network flows are
connecting a multitude of actors in the framework of geographically proximate
area (e.g. micro-region, industrial district), which coordinate the exchange of
resources in pursuit of their own business interest (rational action) but can also
lead to the above-mentioned collateral benefits. By working together, businesses
ensure not only their economic benefit, but attain collective benefits that are
greater than the sum of individual benefits achieved from acting alone. Through
such collaboration, social relationships amongst participants get better (Zhu
et al. 2007).

3.8 The relevance of decision-making process

Recently, numerous networking tools have been made available to individuals
and organizations mainly to help establish and maintain virtual communities.
The common characteristic to all of them is that actors build and maintain

their own social networks based on trust, which are, then, connected to other networks through hubs (individuals that are actors of two or more networks). The aim of this study approach is to create a structure for modelling and enabling the creation of a new trust model such as instances of the proposed approach using hierarchical multi-attribute decision support, social network evaluation – mostly consideration and visualization (Abstractbook 2015).

4 Case studies

Seven industrial symbiotic case studies in this context of modelling are called options or variants. The results obtained are usually subjective and based on the responses of the individual or group's decisions. It is the degree of realization of the decision that must be examined based on the parameters and the decision-making skills. In this study, the author is the decision maker.

From industrial symbiotic cases we will see that industrial symbiosis has grown over the years, including partners and actors from different regions-districts.

Much of the recent analysis found in literature as description has been centred on environmental costs with less attention to benefits. The environmental benefits of industrial symbiosis are quantified by measuring the changes in consumption of natural resources, and in emissions to air and water, through the increased cycling of materials and energy. The economic benefits of industrial symbiosis are quantified by determining the extent to which firms cycling by-products can capture revenue streams or avoid disposal costs; those businesses receiving by-products gain an advantage by avoiding transport fees or obtaining inputs at a discount. In some cases, less tangible benefits in terms of results are obtained from working cooperatively (Chertow and Lombardi 2018, 18).

The brief description of each industrial symbiotic case will help in the evaluation of each case and the comparison between them. The descriptions will clearly show all weaknesses, strengths, and good practices.

4.1 Kalundborg industrial area, Denmark

The industrial symbiotic complex in Kalundborg, DK, is the »influential example of industrial symbiosis in the industrial ecology literature« (Jacobsen 2006, 239–255). »The firms in this industrial park are highly integrated and utilize the waste products from one firm as an energy or raw material source for another« (Industrial, writer in 2018). The Kalundborg project began in 1972. The selection of this case has been made due to its geographic area, used resources, and the used industry sector. The vicinity of the industrial area of Kalundborg enables it to form a network of industries for the exchange of wastes and by-products. When looking at industrial symbiotic organizations, a physical and a social aspect are easily identifiable in Kalundborg. The physical facet-location, diversity of industries, core businesses, presence of the sea, etc., are thought to be hardly possible to model; but on the other hand, the social facet can be modified, improved, and be the subject of work. Thus, the physical facet and, more generally, the geographical settings are considered as the core characteristic of each case (Kalundborg Industrial Symbiosis 2018).

The extent of the material and energy exchanges in 1995 was about 3 million tons a year. The estimated savings totalled 10 million US dollars a year, giving an average payback time of six years (Kalundborg Symbiose 2017; Jacobsen 2005; Jacobsen and Anderberg 2005, 313–335 in van den Bergh, Jeroen and Janssen 2004).

The environmental benefits of industrial symbiosis are quantified by measuring the changes in consumption of natural resources and in emissions to air and water, through the increased cycling of materials and energy. Products gain an advantage by avoiding transport fees or obtaining inputs at a discount (Uzzi 1996, 674–698; Kalundborg Symbiose 2017).

Kalundborg, where interrelations are highly established, is the most often mentioned model of a well-working industrial ecosystem (apart from water-steam loops); there are no cyclical (real) material flows between the collaborating firms (as quoted by Fleig, 2000). Fleig (2000) said that the achievement within the »Kalundborg Symbiosis« is that firms communicated and cooperated so that all the main by-product flows could be directed forward to further users converting them into products, which then leave the system. In the industrial symbiotic organization of Kalundborg, we found the principal firms: Asnaes-Denmark's largest coal-fired power station; an oil refinery owned by Statoil; a pharmaceuticals plant owned by Novo Nordisk; Gyproc, Scandinavia's largest plasterboard manufacturer; and the municipality of Kalundborg, which distributes water, electricity, and district heating to around 20,000 people (Kalundborg Symbiose 2017; Ehrenfeld and Gertler 1997, 3).

The industrial symbiosis of Kalundborg includes nine primary associate firms, some of which are some of the largest firms in Denmark (Kalundborg Symbiose 2017). The firms in Kalundborg are: first, Novo Nordisk, the world's largest producer of insulin. They have about 2,600 employees in Kalundborg. Second, Novozymes, the world's largest producer of enzymes. They employ 500 people in Kalundborg. Third, French-owned Gyproc, which produces gypsum board with around 165 employees in Kalundborg. The municipality of Kalundborg (the fourth partner) handles the water and heat supply for Kalundborg's approximately 20,000 inhabitants. The fifth partner, the Asnaes Plant, owned by DONG Energy, is the largest power plant in Denmark with around 120 employees in Kalundborg (Kaundborg Industries 2017). From the article »Statoil scolded over workers' buyouts« published on 30 March in 2017, we noted that Statoil, Denmark's largest oil refinery, the sixth partner, has around 350 employees (Statoil Scolded over Workers' Buyouts, 2017). The seventh partner is Kara/

Novoren, a waste handling firm that employs 15 people in Kalundborg. The eighth is Kalundborg Forsyning A/S, which supplies the people of Kalundborg with water and district heating and organizes the disposal of wastewater for the entire municipality (quoted in Boshkoska, Džajić and Rončević 2015 from Jacobsen and Anderberg 2005, 221–231; Kalundborg Symbiose 2017).

They have 66 employees in Kalundborg. The ninth member is RGS 90, a Danish soil remediation and recovery firm. A pair of secondary firms, Inbicon and Pyroneer, are owned by the DONG Energy and are part of the network. Inbicon is a lignocellulosic biomass conversion firm, using technology to convert non-food biomass to second-generation bioethanol and other renewable power and biochemical products. The Pyroneer plant specializes in biomass gasification as an endothermic thermal process where solid biomass is converted into gaseous fuel. The produced gas can be used to replace part of the coal used in the existing boiler and the ash can be used in large-scale fertilizer. Included in this dense formation of connections are secondary connections to additional users, such as pig farms, a fish farm, the fertilizer industry, and the cement industry (Behne 2016).

The leading staff at Asnæs Power Station has advanced the notion that environmental impact becomes a downward sloping function of the number of firms cooperating in symbiotic arrangements. This approaches the vision of an industrial ecosystem with limited inputs and limited waste outputs, in which materials and energy are utilized to the full extent possible (AEBIOM 2017). All mechanisms that are assumed to be fostering the sharing of material, services, wastewater, or even organic waste are encompassed in this section; material synergies between firms consist, in most of the cases, in exchanges of liquid and solid waste that can be transported via the help of pipes – out of the 33 projects in Kalundborg, more than half are dominated by water, steam, and heat exchanges, requiring pipes (Kalundborg Symbiose 2017). These pipes usually require resources when being implemented and maintained, but also space and special commodities. The compliance with urban local plan authorities giving permissions for such construction is a mechanism that can facilitate the synergies implementation. It can be a barrier if some regulations are made to limit the possibilities of having the pipes created. In Kalundborg, the municipality has seen cooperation among firms as an opportunity for the town, and they have seen industrial symbiosis being made on an industrial level with the municipality. Without support from the municipality, constructions would have been impossible to carry out (Kalundborg Symbiose 2017).

Since Kalundborg is a town with a relatively short distance between firms, that allows greater possibilities in constructing infrastructures and finding new synergies. Long distances between firms can have an influence when making decisions. Because of a loss of energy, synergy projects are distance limited (Kalundborg Symbiose 2017), according mainly to the nature of the material exchanged. There may also be an economical reason to distance limitation. Since pipelines are expensive, a cost-benefit analysis defines the economical break-even distance point up to where it remains economically profitable to have a synergy. Therefore, the synergy distance may vary according to the value of the material exchanged, the cost of infrastructures, the return rate of investment, but also its payback time – usually 3–4 years' payback time has been observed in Kalundborg (Kalundborg Symbiose 2017).

An affinity and the need to establish and maintain links between firms can appear in a town with a size-limited industrial area. In Kalundborg, intimacy plays an important role from the beginning of synergies (Kalundborg Symbiose 2017). Maybe some synergies could have started because most of the chairmen of the firms shared common social activities (golf and social clubs) and interacted with each other. These interactions help the development of industrial symbiotic networking by overcoming confidentiality issues.

In the western European culture of entrepreneurship, there is a strong sense of independence of the private sector. In that sense, the industrial symbiotic system in Kalundborg is almost totally driven by market forces, where public governance has had very little impact on the industrial symbiotic development (Kalundborg Symbiose 2017). This reactivity to economic incentives produced by the market gives the industrial symbiotic system a dynamic and efficient development. If firms, as important actors within industrial symbiosis, feel forced to produce a non-efficient outcome, their willingness in doing so will most probably decrease. This can lead to a slowdown of the industrial symbiotic development process (Kalundborg Industrial Symbiosis 2018).

In industrial symbiotic networking, when collaborations between a high number of actors are developed, there may be an emergence of a possible gap in terms of communication or other organizational aspects. Therefore, it is important to tackle this problem by creating a common communication platform between active firms. In Kalundborg's case, this led to the creation of the industrial symbiotic institute which aims is to accelerate and increase the complexity of the exchanges. This leads to a better atmosphere of cooperation between the actors, which is a key element in the introduction and development of industrial symbiosis. These relationships were incontestability in terms of the role of governance in industrial ecology projects (Gallaud and Laperche 2016).

Additional estimates from Kalundborg suggest that a 15 million dollars' collective annual savings has been achieved, primarily on resources, from a total investment of 90 million dollars. Total savings through 2002 are estimated at 200 million dollars, but a detailed analysis to support this conclusion has not been performed (Christensen 2004). The achievement of high levels of environmental and economic efficiency has also led to other collective benefits involving personnel, economic/technological advancement, and information sharing. The critical lesson from the Kalundborg symbiosis concerns the evolution of exchanges within a network – the Kalundborg case was not centrally planned and did not unfold all at once but evolved over the last forty years as a series of bilateral contractual arrangements between firms (Christensen 2004). Therefore, an economic balance is made between the revenue streams from waste exchanged and the costs of construction and the maintenance of infrastructures. »The industrial symbiotic partners have an economic advantage because all agreements are based on solid business principles« (State of green 2017).

Industrial symbiosis involves economic benefits for the involved social actors, e.g. savings in production cost, savings in the purchase of new materials, savings in the management of secondary resources, and additional revenue from re-processing secondary resources into by-products and primary resources (Ehrenfeld and Gertler 1997; Doménech 2010; Doménech and Davies 2011; Ehrenfeld and Gertler 1997).

4.2 Aalborg industrial area, Denmark

The industrial area of Aalborg has the potential to cover our needs in terms of the experimentation of the second best industrial symbiotic case in the evaluation. Even though most of the traditional activities in Aalborg, Denmark, have been offshored abroad for (mainly) cost salary reasons, heavy manufacturing activities still characterize the city economy such as the cement industry, a power production plant, and a harbour. Synergies of the industries in Aalborg happen within a radius of 1–2 km distance between firms (Aalborg Portland A/S 2010, Aalborg Industries 2015, 2017; Aalborg Portland Miljøredegørelse 2012, 2017; Aalborg Kommune 2010, 2017) in north Denmark. Aalborg is a mid-sized town that gathers all the prerequisites present in most of the industrial symbiotic cases around the world: it has an opening on the sea, harbour activity, an important industrial area, manufacturing businesses, etc. However, very few symbiotic partnerships between firms are in place (Scrase et al. 2009; Aalborg-Renovation-Groenne-Regnskaber 2018).

There are only a few productions of natural resources in Aalborg. One of them is chalk, extracted nearby Aalborg and used in cement production at Aalborg Portland (Aalborg Kommune 2017; Aalborg-Renovation-Groenne-Regnskaber

2018; Aalborg Supply 2018). However, the price of the water may have an influence on firms' efforts to lower their consumption. Today, Aalborg Portland is a part of a symbiotic partnership (Aalborg Supply 2018). Elsam A/S, Nordjyllandsværket sends 30 tons per day of chalk slag to Aalborg Portland that sends, in turn, desulphurization gypsum, which is used to clean the extraction of chalk slag (Aalborg Supply 2018; Årsrapport-Aalborg Portland A/S 2013; Aalborg Portland Miljøredegørelse 2011).

By using the chalk from the power plant, the amount of resources extracted is minimized. Aalborg Portland is already recycling wastewater in an internal circulation. In 2009, 347,230 tons of water was recycled internally. In Aalborg 26,779 gigajoules have been used for district heating (Aalborgportland Portland Environmental Report 2011, 2013). Most of the waste is recycled, but it is not considered as an industrial symbiotic synergy since the waste is going to other firms that are not interconnected. Reno Nord uses garbage from households and firms to generate electricity and district heat thanks to a combined heat and power plant. In 2009, 186,879 tons of garbage was treated at Reno Nord (Nord n.d.). The outputs of the process are heat and electricity. The sludge from the incineration process is sorted, some parts are used for asphalt, and others for metal recycling; in this way, it is a resource for other firms whose necessity of virgin resources will decrease.

»Due to the limited access to raw natural resources, environmental taxes have given incentives to the firms in recycling waste and minimizing new resources. Using waste products of another industry does not incur any extra fee« (as quoted in Almasi et al. 2011, 38).

Urban local plans exist in most areas of Denmark. The »Law of planning« (as quoted in Almasi et al. 2011) acts as a guideline to design them. The local plan dictates what a specific area is used for, where pipes and electricity should be placed, and so on (Aalborg Kommune 2011; Aalborg Supply 2018).

»Environmental regulations give incentives to develop environmental friendly alternatives in Aalborg. However, some environmental taxes may induce negative impacts on the development of industrial symbiotic synergies. Because of the industrial diversity in the Aalborg East area, a wide range of technical and innovative processes are taking place« (Almasi et al. 2011, 38).

Some firms, such as Siemens Wind Power, Reno Nord and Fibertex have been investing in research and development to come up with new technologies that could be applied to the production process. This often results in productivity gain. However, technical progress has also allowed the development of a wider range of possible synergies. Since some synergies in Kalundborg or Kwinana were not possible before, mostly due basic or insufficient technology that suffered important energy losses, technological advances have now made it

possible to slow down the high entropy transformation of energy. Therefore, a single flow
of energy, hot water for example, can now go through several uses at several places while
limiting the heat loss. Some other techniques of waste conditioning that did not exist before
also allow the reuse of a wider variety of waste and, thus, enhance the synergies develop-
ment, like gas desulphurization and others (Almasi et al. 2011, 40).

In Aalborg East, the industrial area is focused upon 3.6 km^2, assembling the main industries of the city. To define the degree of physical proximity between firms, a mean distance between firms is measured. The greatest mean distance possible, meaning the distance between the farthest located firms in the industrial area, is taken into account. »However, in order not to bias the average result, Aalborg Portland has been discarded. Indeed, even in the case where Aalborg Portland is involved in a synergy, the solid nature of inputs needed indicates that the trans-action would probably be achieved through truck transportation, as is actually the case concerning the exchange of fly ash between the latter and Reno-Nord« (Almasi et al. 2011, 39). The fact that Aalborg East has a diversified industrial structure, with some of the firms using advanced technologies in their produc-tion process, indicates that some of the present actors possess knowledge and engineering capacities of implementing advanced technology-based synergies, if and when required.

»Among firms specialized in different manufacturing or service based activ-ities, the presence of specific infrastructures, such as the wastewater treatment plant of Aalborg East, may facilitate interconnections between firms flows. »The diversity of core businesses develops the possible combinations in terms of by-products and wastewater complementarity«. Thus, one firm's waste has a greater chance to become another's raw material« (Almasi et al. 2011, 40).

Studies (Doménech 2010) in this area examine how shared beliefs, values, and norms develop within a social system and how these, in turn, influence an organization's behaviour and function. Industrial ecosystems may constitute new organizational fields that are based on geography, compatible material flows and coordinated resource management rather than industry classification. Economic geography examines why industries tend to concentrate in regions and measures the resulting economic benefits to firms and the regions overall (Scott 2000).

These benefits include improved access to factors of production and reduced costs through economies of scale. Knowledge spillovers and innovation are thought to result from frequent inter-firm communication and cooperation due to proximity. There are many types of successful regional industrial systems. Among these are diverse urban economies, clusters dominated by a few related industries, districts consisting of small- and medium-sized enterprises that coop-erate to innovate, and satellite districts that house subsidiaries of multinational firms. Regional systems are thought to evolve from locations where collocated firms are unconscious of each other and simply benefit from economies of scale to systems that include dynamic learning and coordination to boost regional competitive advantages. Within regional economies, personal relationships (social ties) are thought to provide non-economic incentives for managers to cooperate in their mutual interest. Industrial ecosystems can also be considered as a type of regional economy (Kalundborg Symbiose 2017). Collaboration in such systems centres on material exchanges and resource management issues with there being an awareness of the resulting public environmental benefits (Chertow, Weslynne, and Espinosa 2008). Social forces operate within both regional economies and organizational fields and are thus worthy of study to understand how inter-firm collaboration takes place within industrial ecosystems (Jarvis 2008).

The feasibility of such collaboration must be achieved with the support of a technical and also cost-benefit analysis. The major part of industries operating in Aalborg are using input materials that are present in a finite quantity in the environment (freshwater, steel, gypsum, glass fibres, etc.), or, at least, non-renewable in a human lifetime span. Therefore, their availability decreases and their prices increase. It supports the idea aforementioned and encourages industrial symbiotic development.

A common communication platform, which can be assimilated as a coordinating body, would facilitate partnerships between firms having technical compatibility, circulate information flows throughout the area, and organize workshops. It is not possible to assess the degree of integration of such an entity since no industrial symbiotic network has been yet developed in Aalborg East. However, it is still possible to encourage such an initiative in the near future. It is thought to enhance the development of industrial symbiotic synergies in an already existing industrial area (Kalundborg Symbiose 2017).

Some firms are already interacting in some way, but other than in a usual business context, cooperating with each other, mainly regarding waste disposal transactions.

4.3 Guayama industrial area, Puerto Rico

Several conditions converge to make Puerto Rico of great interest to industrial ecologists. The island presents an intricate mix of intensive industry and some very diverse ecological systems within a bounded geographic space that is small enough to be carefully studied but large enough to require multifaceted solutions that are useful and comparable with many other situations around the world (Chertow 2004). Puerto Rico is a commonwealth of the USA located in the Caribbean Sea with a total land area of approximately 9,000 km^2 and a population of approximately 4 million people. It is a world leader in pharmaceutical manufacturing, producing 16 of the top 20 best-selling drugs in the USA (Chertow, Weslynne, and Espinosa 2008) and is also strong in the manufacturing of electronics and medical devices. The notion of business clusters has played a major role in Puerto Rico's economic development since the mid-20th century, through targeting particular industries as well as organizing industrial parks with the infrastructure to facilitate their operations. The island currently has 139 parks managed by the Puerto Rican Industrial Development (PRIDCO) varying in size, number of industries, industry origin (local or multinational), and industry type (Eilering and Vermeulen 2004; Melia 2007).

As described by Chertow and Lombardi (2005, 6535–6541): »an evolving network of inter-firm exchanges in Guayama, Puerto Rico provides a tractable illustration that can be used to analyse the environmental and economic case for industrial symbiosis«. Puerto Rico is a commonwealth of the USA, thus sharing most laws and business practices from the USA. The municipality of Guayama, on the southeastern coast of Puerto Rico, measures 169 km^2 and has a population of approximately 42,000 people. Before 1940, Guayama was primarily an agricultural economy with some light manufacturing. After a period of light industrialization in the 1940s and 1950s, the current industrial profile began developing.

In 1966, Phillips Petroleum opened a petrochemical refinery, which today is owned by a partnership between Phillips and Chevron. In the 1980s several pharmaceutical firms opened manufacturing plants in the Jobos Barrio (neighbourhood): Baxter (1981), Wyeth (1985), IPR Astra Zeneca (1987), IVAX (pre-1994). There are also light manufacturing businesses in the industrial zone: Lata Ball (aluminium can be manufacturing); Alpha Caribe (plastic bottle manufacturing); PR International (heavy machinery repair); and Colgate Palmolive (oral care and detergents manufacturing). In November 2002, AES Corporation brought online a 454 MW coal-fired power plant with atmospheric circulating fluidized bed (ACFB) technology (Chertow and Lombardi 2005, 6536).

Guayama hosts many industries: a fossil fuel power generation plant, pharmaceutical plants, an oil refinery, and various light manufacturers. Current exchanges in Guayama include the new AES coal-fired power plant using reclaimed water from a public wastewater treatment plant (WWTP) for cooling and providing steam to the oil refinery. Additional steam and wastewater exchanges are under negotiation between next-door pharmaceutical plants, the refinery, and the power plant. Beneficial reuse of the coal ash has begun as a means of stabilizing some liquid wastes. An analysis follows of existing and potential exchanges in Guayama with the power plant and refinery as critical participants. Costs and benefits to each firm are explored as well as regulatory, political, and other relevant factors. From its founding in 1941 until the 1990s, the Puerto Rico Electric Power Authority (PREPA or AEE by its Spanish acronym) built and owned essentially all of the power generation and distribution facilities on the island. PREPA has maintained control of electricity distribution; however, in the face of United States' energy restructuring and increasing need for power generation in the 1990s, PREPA opened this area to proposals from independent power generators. In 1978, a federal law was passed in the USA to encourage the use of renewable, non-polluting methods of power generation (Melia 2007). The Public Utilities Regulatory Policies Act (PURPA) requires a public utility (such as PREPA) to purchase electricity from small producers and from independent producers that maintain qualifying status if the price of the electricity is below the utility's own marginal costs of production. For an independent generator to maintain »qualifying facility« (QF) status, the facility must use at least 5% of its energy output for products other than electricity: steam and desalinated water are common cogeneration products (Chertow and Lombardi 2005, 6535–6541). When PREPA opened to proposals from independent generators, it required that all proposed facilities qualify as cogenerators under PURPA. In the 1990s, two independent projects were approved as QFs: AES Guayama and Eco Electrica, a natural gas combined cycle plant in the south western part of the island producing electricity and 2 million gallons/day of desalinated water (Chertow and Lombardi 2005, 6535–6541).

»A coal-fired power plant, owned and operated by the AES Corporation, draws five million gallons/day of process water from nearby sources thus avoiding freshwater withdrawals and, through steam sales, significantly reduces emissions from a nearby refinery« (Chertow and Lombardi 2005, 6535).

This pattern is looking to be repeated in Guayama based on several requests, particularly for the exchange of chemicals, now being evaluated by nearby firms. In essence, the successful initiation of trading within co-located firms appears to bring a shift in thinking, a change in the dominant trajectory of firm individualism, creating a willingness to consider further trading. Given the nature of environmental and economic benefits identified in the case of Guayama with respect to air, water, and by-products, it is reasonable to infer that these benefits, none of which are very unconventional, can be found in many other comparable situations.

In the authors' opinion, the organizational hurdles may be a lack of information across firms, a perception of high transactions costs across firms, insufficient trust or communication across firms, or a lack of a regulatory push. Key, then, is catalysing the early exchanges so that firms will be open to the potential benefits of additional ones when economically and environmentally desirable (Boshkoska, Džajić and Rončević 2015).

Yet, as in the case of Guayama, regional government action to foster waste-water reuse or national government action to encourage cogeneration as in PURPA can also be powerful catalysts. Through selected policy interventions such as those described to reward wastewater reuse or additional use of waste heat, government action could advance symbiosis which, in turn, could bring additional public and private benefits in its wake. The private benefits of these inter-firm transactions are high: the availability of steam, for example, has a value greater than 8 million dollars/year (Chertow and Lombardi 2005, 6535–6541). In addition to the quantifiable economic benefits of the symbiosis activity, environmental benefits of steam sharing were also found to be substantial resulting in a 99.5% reduction of sulphur dioxide (SO_2) emissions, an 84% reduction in nitrous oxide (NO_x), and a 95% reduction of particulate matter smaller than 10 mm. Indeed, were it not for the ability of the coal plant to use wastewater, siting of the plant would not have been possible in this dry part of the island. Thus, in addition to agglomeration benefits, an effect of the industrial symbiosis was to facilitate regulatory permitting, thus helping the power station to achieve its license to operate (Chertow and Lombardi 2005, 6535–6541). Although symbiotic linkages have been in place for only a short time in Guayama, further opportunities for by-product exchanges are already being considered for power station ash and for wastewater with a neighbouring pharmaceutical firm. By sending its pre-treated wastewater directly to the power station, the pharmaceutical firm would avoid water discharge fees and earn additional revenue. The power station, in turn, would benefit from having a cheaper water source. The environmental benefits of these potential exchanges include an overall reduction in virgin material use, the avoidance of wastes being discarded, and a reduction of water consumption. Further research will determine if the continued learning about the economic and environmental benefits of industrial symbiosis will lead to a stage of dynamic environmentally related agglomeration economies in Guayama. Interestingly, there is some social context behind these exchanges, for example, with regards to social networks, the willingness to consider new opportunities, the integration of the community, rules, and the division of labour.

As we may see from the literature, in 1982, halogenated solvents were identified in wells located between the refinery and pharmaceutical facilities. The US Environmental Protection Agency (USEPA) placed the wells on its National Priority List of hazardous wastes contaminated sites designated for clean-up (The International Journal of Life Cycle Assessment 2015). »The process of identifying the responsible parties and devising a clean-up plan served to bring managers in this area together. While the circumstances were not pleasant, they facilitated familiarity among the key players. Subsequently, there has been movement of managers between firms, as well as joint committees that work on common concerns such as emergency planning« (N. Marquez, Loss Prevention Manager, Chevron-Phillips Core Guayama, Puerto Rico, personal communication, 2005 quoted in Chertow and Weslynne, 2009, 128–151). Aspects such as stronger social bonds and a willingness to cooperate seems to play an important role in the development of synergies, although they might not be sufficient for the symbiosis to realize.

One piece of the research about industrial symbiotic networks that originated in Guayama, which focused on the participants and the community, was conducted by Chertow and Lombardi (2005, 6535–6541). It quantifies the economic and environmental costs and benefits to participants and concludes that there are substantial benefits, although they are irregular in the participating organizations. In Guayama, we saw that when policy intervention can be seen as a viable means to motivate the most common occurrences of exchanges of resources between groups of firms, industrial symbiotic network has been identified.

4.4 Barceloneta industrial area, Puerto Rico

Barceloneta and its neighbouring municipalities, Arecibo and Manati, are often referred to as having one of the highest concentrations of pharmaceutical manufacturing facilities in the world. The island's pharmaceutical industry, which still produces 13 of the 20 best-selling drugs in the USA, gained dominance in the 1970s with the help of US incentives. It accounts for a quarter of the island's gross domestic product, with 36.5 billion dollars in annual exports (Melia 2007). Currently, there are many pharmaceutical facilities operating in and around Barceloneta, an area with historically abundant groundwater supplies. These services are in the manufacturing stage of pharmaceutical fabrication, including the chemical, biological synthesis of active ingredients, and the preparation of final goods. The services benefit from static accumulation as a large pool of capable and semi-capable labour as a basic shared infrastructure (roads…).

> *The participation of firms is in utility sharing, joint service provision and by-product exchange. Instances of industrial symbiosis in practice in this cluster include a shared 38,000 m3/day capacity wastewater secondary treatment facility built primarily for the treatment of pharmaceutical wastewater (industry specific utility sharing) and financed by the firms. The firms benefit from capacity allowance guarantees that lets them increase*

their production without increasing their wastewater treatment cost (B. Martir, Principal, Bemar Associates, Barceloneta, Puerto Rico, personal communication 2005 quoted in Chertow, Weslynne and Espinosa 2008, 1307).

Sludge from the plant is converted into a fertilizer that is applied to an adjacent hay farm where 68,000 kg of food are harvested annually and sold as animal feed (by-product exchange). Waste management firms perform closed loop solvent recovery for several pharmaceuticals, reducing the latter's virgin material use, and purchase and transportation costs. Waste brokers also facilitate occasional sales of used and off-spec materials (industry-specific joint service provision) and other manufacturers (P. Sanchez, owner/ Director, Waste Exchange, Trujillo Alto, Puerto Rico, personal communication, 2002 quoted in Chertow, Weslynne and Espinosa 2008, 1307).

Some industrial groupings of firms, as in Australia's mineral processing industry and Austria's recycling network in Styria, find productivity-enhancing efficiencies through self-organization. Stronger regulatory roots combined with elements of self-organization are seen in those uncovered in Puerto Rico. Barceloneta pharmaceutical firms teamed up in a private initiative to share wastewater processing in response to regulatory changes and, subsequently, further exchanges developed (Ashton 2008; Boshkoska, Rončević and Džajić 2018).

Despite food, chemical, packaging, electronic equipment, metal fixture manufacturers, and waste management firms, there are global leaders such as Abbott, Bristol-Myers Squibb, Merck, and Pfizer. Beginning in the 1950s, Puerto Rico, a US commonwealth territory, promoted generous tax benefits and low workforce costs relative to the US continent to attract manufacturing firms. Eight regional firms participate in the Barceloneta Wastewater Treatment Corporation Advisory Council, a joint agreement, started in 1978, to construct and oversee operation of the region's secondary wastewater treatment plant, which handles 31,415 m³/day (8.3 million gallons/day). The plant is owned and operated by the Puerto Rico Aqueduct and Sewerage Authority. Advisory Council members provided 70% of the plants' operation and maintenance costs, as well as technical assistance to the authority (Ashton 2008). The pharmaceutical services were financed to manage the wastewater treatment plant initiative. Managers in the participating firms met frequently to discuss common problems; initially these were focused on the wastewater plant, but later evolved to include other shared resource constraints. The Advisory Council has served to institutionalize cooperative resource management practices among the firms and deepen the social embedding of inter-firm relations (Corder et al. 2014; Granovetter 1985).

These on-going synergies were made possible through the shared positive experience of the Barceloneta Advisory Council, which is made up of the pharmaceutical facilities that financed and help manage the wastewater treatment plant initiative. Managers in the participating firms meet frequently to discuss common problems; initially these were focused on the wastewater plant, but they later evolved to include other shared resource constraints (C. Bassat, Director, Community Affairs, Merck, Sharpe and Dohme-Puerto Rico, Arecibo, Puerto Rico, personal communication quoted in Chertow, Weslynne and Espinosa 2008, 1308).

Renken et al. (2002) and Rivera-Santos (2002) define that the Barceloneta cluster also highlights an example of agglomeration dis-economies. While abundant water supply initially encouraged the spontaneous co-location of the pharmaceutical firms. In recent years, the groundwater reservoir has been increasingly threatened by high extraction rates and localized areas of contamination, which is a negative externality of co-location. In-house programmes for reducing water consumption and water cascading were implemented by a few Barceloneta firms to confront this issue (Ashton 2008), but cascades across firms have not been explored for this purpose. This unrealized opportunity of industrial symbiosis could counteract the agglomeration dis-economies in Barceloneta (Chertow, Weslynne and Espinosa 2008).

Ashton (2008), for example, observed a positive correlation between trust and involved actors in the social hierarchy in industrial symbiotic linkages in Barceloneta, Puerto Rico. She analysed the different strategies, to generate trust and commitment and engage local communities. The conclusions of the research emphasize social relations as a key element in the development of industrial symbiosis, over system engineering or industrial-technological solutions. She found that trust is mostly generated on a face-to-face basis, through such aspects as steering committees or personal linkages. The creation of trust requires a considerable amount of time and effort to build. Fundamentals in building the relations and developing common project are proximity and engagement in the local context (Doménech 2010).

4.5 Kwinana industrial area, Australia

Kwinana compares favourably with well-known international industrial symbiotic cases in terms of the current level and maturity of industry involvement and collaboration and the commitment to further explore regional resource synergies. »Kwinana stands out with regard to the number, diversity, complexity and maturity of existing synergies« (van Beers et al. 2007b, 65). A lot of different local synergy opportunities still appear to exist, mostly in three broad areas: water, energy, and inorganic by-product reuse. The Centre for Sustainable Resource Processing (CSRP), to improve the further development of new regional synergies, a joint initiative of Australian minerals processing firms, research providers, and government agencies, has undertaken several collaborative projects. »These include research to facilitate the process of identifying and evaluating potential synergy opportunities and assistance for the industries with feasibility studies and the implementation of selected synergy projects« (van Beers et al. 2007b, 55).

The Kwinana Industrial Area (KIA) was established in the 1950s and is Western Australia's most significant heavy industrial region. The area of 120 km^2 is located approximately 40 km south of Perth. Western Australia is the largest and most sparsely populated state in Australia. The state has rich endowments of natural resources, including, but not limited to, iron ore, bauxite, gold, nickel, mineral sands, diamonds, natural gas, oil, and coal. Heavy process industry is concentrated in a few industrial areas, of which the Kwinana Industrial Area is by far the largest and most diverse. About 3,600 people work in the area's core industries and many more in related sectors and service jobs. The total economic output of the area exceeds 4.3 billion dollars annually (Merz 2002). The KIA is home to a diverse range of industries ranging from fabrication and construction facilities to high-technology chemical and biotechnology plants and large resource processing industries, such as titanium dioxide pigment production and alumina, nickel, and oil refineries. The instance of industrial symbiosis has considerably increased since the late 1980s, providing economic, environmental, and social benefits to the firms involved, the nearby communities, and the State. Following the successful development and start-up of two cogeneration plants in 1997 and 1999, the KIC (KIC – the local industry group) decided to explore further opportunities for industrial symbiotic exchanges in the area (Synergies-Kwinana Industries Council 2018). A regional economic impact study was conducted and included an analysis of the principal material and energy flows within the area and the level of industrial integration (Merz 2002). Between years 1990 and 2000, the number of existing interactions increased from 27 to 106 (including 68 between core process industries and 38 with service and infrastructure industries) (Synergies-Kwinana Industries Council 2018).

Heavy process industries dominate the KIA. These include (Merz 2002) a 2,000-kiloton (kt.)/year (yr.) alumina refinery (Alcoa), a 70-kt/yr. nickel refinery (Kwinana Nickel Refinery), a 105-kt/yr. titanium dioxide pigment plant (Tiwest), 850-kt/yr. lime and cement kilns (Cockburn Cement), a 135,000-barrel/day oil refinery (BP-Kwinana Oil Refinery), and an 800-kt/yr. pig iron plant (HIsmelt). These are complemented by a variety of chemical producers, including CSBP, an Australian fertilizer and chemical firm based in Kwinana, Western Australia (Industry involvement CSBP 2018) (ammonia, ammonia nitrate, cyanide, chlor-alkali, and fertilizer plants); Coogee Chemicals (inorganic chemicals); Nufarm (herbicides and other agricultural chemicals); Nufarm Coogee (a chlor-alkali plant); Bayer (agricultural chemicals); Chemeq (veterinary products); and Ciba and Nalco (water treatment and process chemicals). Moreover, there are important utility operations, including two power stations (900 MW coal-, oil-, and gas-fired and 240 MW combined cycle gas) both owned by Verve Energy; two cogeneration plants [respectively 116 MW (Kwinana Cogeneration Plant) and 40 MW (Verve Energy)]; two air separation plants (Air Liquide and BOC Gases); a grain handling and export terminal (CBH); port facilities (Fremantle Port Authority); and water and wastewater treatment plants (Water Corporation). Historically, considerable supply chain integration has occurred between these industries in the area (van Beers et al. 2007a; Synergies-Kwinana Industries Council Strategic Plan 2005–2008 2018).

In 1991, the core industries were established by the Kwinana Industries Council (KIC) (Corder 2008; van Beers et al. 2007b, 55–72). The KIC seeks to foster positive interactions between member firms, government, and the broader community. Based on the desire of the local industries and the state government to understand and document the full economic and social contributions of the KIA, the KIC initiated regional economic impact studies, which included an analysis of the principal material and energy flows within the area and also assessed the level of industrial integration (van Beers et al. 2007b, 55–72; Synergies-Kwinana Industries Council Strategic Plan 2005–2008 2018).

> *The recently updated inventory of existing synergies showed that these are quite diverse in Kwinana. 47 synergy projects are in place, including 32 by-product synergies and 15 utility synergies. The inventory was compiled from industry surveys and is therefore not necessarily complete. Each synergy can involve multiple material, water, and/or energy flows* (van Beers et al. 2007b, 76).

Kwinana industrial symbiotic area presents numerous specific facilities that, as an effect, are believed to foster the diversity and the feasibility of synergies between firms: two water and wastewater treatment plants and two cogeneration plants. Kwinana believes in the importance of intimacy in the development of industrial symbiosis. The limited competition between operating firms and the isolation of other major industrial centres in Eastern Australia are believed to be a consequence of such social proximity between these actors. The proximity of manufacturers that can provide specific equipment such as pipelines, boilers, and all necessary infrastructures in order to achieve exchanges, is considered as an important mechanism. Bladt Industries, Aalborg Industries or Aalborg Energie Technik is one of these firms (Bladt Industries n.d., Aalborg Industries n.d. and Aalborg Energie Technik A/S n.d.), both located in Aalborg that could provide with such products. There is a strict regulation with regards to handling organic materials in order to produce alternative fuels (Synergies-Kwinana Industries Council Strategic Plan 2005–2008 2018; Christensen 2004; van Beers et al. 2007b, 55–72). The Kwinana Industrial Area (KIA) provides a strong foundation for the identification and development of synergy opportunities. The 2002 Economic Impact Study identified over 100 interactions between existing industries in the KIA. Put differently, the degree of diversity of synergies would be positively correlated to the diversity of industry sectors (Synergies-Kwinana Industries Council Strategic Plan 2005–2008 2018). The case of Kwinana gives a great deal of importance to the presence of a coordinating body as a means to foster interactions between member firms.

Another driver for synergy development has been staff mobility. Mobility of staff between neighbouring operations has contributed to synergies in two different ways. In the case of staff mobility between different industries, for example in Kwinana, it amounted to a greater awareness of industrial operations and their associated process inputs and outputs, which has contributed significantly to identifying synergy opportunities (Van Beers et al. 2007a, 830–832).

Kwinana is increasingly subject to urban encroachment and resulting higher community expectations with regard to environmental and safety performance and overall amenity. Kwinana is situated on the coast of the Cockburn Sound, a sensitive marine environment and recreational area for local residents (Synergies-Kwinana Industries Council Strategic Plan 2005–2008 2018; van Beers et al. 2007b, 55–67).

The chance to transfer the discharge of treated process wastewater from the coastal area into the deep ocean outlet as part of the Kwinana Water Reclamation Plant (KWRP) was therefore an important consideration for CSBP (Industry Involvement, CSBP 2018), Tiwest, and BP to purchase the higher-cost water from KWRP. Moreover, the CSBP chemical and compost plant built in 2004, an innovative nutrient-stripping wetland served to further reduce the nitrogen discharges to the adjacent Cockburn Sound. The pilot wetland was constructed on land leased from the BP refinery. The wetland was planted with sedges and incorporates a number of biological processes that will reduce the level of nitrogen in the CSBP's effluent stream. Some of BP's effluent is also released into the wetland and it is found to provide additional benefits by supplementing the carbon loading. The core business focus of the participating industries sometimes functioned as a synergy barrier. In Kwinana (and aslo in Gladstone), the emphasis of site personnel is on core business activities, resulting in potential missed synergy opportunities unless there is an overwhelming commercial benefit. This is recognized by various site personnel, who see one of the main aims of the regional synergies research as being the identification and progression of synergy opportunities that are unrelated to the core business (van Beers et al. 2007b, 68).

Major capital projects served as a synergy driver and trigger, particularly with respect to new operations or significant capacity expansion projects in existing operations. In Kwinana, two new industrial facilities were built and commissioned in 2004 (the Kwinana Water Reclamation Plant and the HIsmelt direct-reduction iron-making plant). The HIsmelt plant is able to source a number of inputs locally in the Kwinana area, such as lime, lime kiln dust, treated wastewater, and provide outputs with the potential for reuse in Kwinana, such as slag and gypsum. HIsmelt triggered the undertaking of the Kwinana Water Reclamation Plant (KWRP), as the groundwater allocation for the area had already been licensed to the existing industries, and there was limited availability of catchment (potable) water in the metropolitan area of Perth. The distance between firms was a barrier to synergy development, the distances still pose a challenge with regard to, for example, the recovery and reuse of process energy and water.

Kwinana firms experienced obstacles in getting governmental approval for the use of alternative fuels and raw materials. Although some by-product synergies appear technically and economically feasible and had a positive sustainability impact (e.g. alternative fuels in cement kilns and the use of bauxite residue for soil conditioning), their practical implementation had been halted by uncertainties in the legislative structure, in particular with regard to the final responsibility for approved reuse options and community concern. Additionally, if a by-product is classified as a controlled waste (for example, fly ash), strict transportation procedures and requirements apply. Regulations could also be a trigger for regional synergies (Corder 2008).

> Another synergy driver was the technical obsolescence of existing process equipment. The Kwinana Cogeneration Plant is located on land of the BP oil refinery, produces all process steam for the refinery, and generates electricity for BP as well as the grid. The cogeneration plant is fired with excess refinery gas from the oil refinery supplemented with natural gas. The cogeneration plant, built in 1996, took the place of the BP steam boilers, which were in need of replacement at the time. This synergy allowed BP to decommission its old inefficient boilers, estimated to have saved the refinery in the vicinity of 15 million dollars in capital expenditure, while ensuring a cost-competitive, reliable source of steam and electricity for their refinery. Moreover, the refinery has achieved greater refinery process efficiencies as a result of the greater and more flexible availability of high-pressure steam from the cogeneration facility. Major Brownfield expansion functioned as a synergy trigger as well (van Beers et al. 2007b, 67).

In this industrial symbiotic case, it is seen that the focus is on the process of transmission, which is most relevant to the diffusion of industrial symbiotic networks. The researchers in Kwinana work closely together with the local industries; the type and level of assistance to the development of new synergies depends entirely on the specific research needs of the involved firms (van Beers, Bossilkov, and van Berkel 2006).

4.6 Gladstone industrial area, Australia

Gladstone, situated on the central Queensland coast about 540 km north of Brisbane, has a population of about 40,000 in the region which extends from Boyne Island in the south to Yarwun in the north. The Gladstone area has been strategically earmarked by the Queensland State Government for significant future industrial development by large industry (Corder 2008). These industries are as follows: Queensland Alumina Limited – described as the largest alumina refinery in the world (Bossilkov, van Berkel and Corder 2005). It has the capacity to process annually 8 million tons of bauxite which is used as a raw material to produce 3.7 million tons of alumina; NRG-Gladstone Power Station – described as the largest coal-fired power station in Queensland; Cement Australia – receives limestone mined from the nearby East End Limestone Mine; Boyne Smelters Limited – produces over 500,000 tons of aluminium annually; Orica Chemicals Plant Limited – currently produces 275,000 tons of ammonium nitrate, 50,000 tons of sodium cyanide, and 9,500 tons of chloride annually (Bossilkov, van Berkel and Coreder 2005); Comalco Alumina Refinery (CAR) – described as the first new alumina refinery in Australia since 1995 (Bossilkov, van Berkel and Coreder 2005); Central Queensland Port Authority – handles about 40,000,000 tons of coal/year (Bossilkov, van Berkel and Coreder 2005); Queensland Energy Resources Limited (QERL); and Gladstone Area Water Board (GAWB) which supplies water to industries and community within the Gladstone region (van Beers et al. 2007b, 62).

The Gladstone State Development Area is a 14,000ha land bank that has been specifically reserved for large-scale resource processing, metals smelting, and downstream manufacturing industries. Main industrial operations have been part of the Gladstone region since Queensland Alumina Limited commenced its operation in 1967. Other industries located within the region include a major power station, cement manufacturing, aluminium smelters, plus chemical manufacturing and large port facilities, and utilities providers. There are nine major industries operating in the region that form an association called the Gladstone Area Industry Network (GAIN), which also collaborates with the Gladstone City Council (Corder 2008, 20–40). The improvement of the initiation of symbiosis was influenced by factors such as the population number, the location and size of the area, and the diverse industries located in an industrial area. For example, the small community population of 20,000 and the small number of industries in a limited area. It was also confirmed by the Symbiosis Institute that the »short« mental distance made the informal connections necessary for developing trust between the industrial community as well as the local community. In Gladstone, a relatively large industrial area away from the city with small number of industries may have contributed to the comparatively slow contributions to symbiosis by communities and governments (Corder et al. 2014, 340–361).

From the literature, it was clear that connection, communication, and collaboration was important to bring internal and external partners together to successfully implement industrial symbiotic projects in the various cases. There is a connection between the levels of social capital needed between different partners to reach results, such as sustainability. Such symbiosis is much more than by-product exchanges between industries co-located or physically close to each other. It was identified that it is the result of the developed relationship between the partners of different industries involved and also with communities and governments closer to industrial area, which enables the trust that is essential to allow the sharing of information to occur. These include both risks and benefits equally, thus reducing the barriers such as a lack of sharing information, plus risks and benefits due to commercial confidentiality.

In one of the projects that was made on regional synergies in Gladstone, it was found that: more regional synergies in Gladstone were not taken up because there were not enough financial benefits or external drivers to justify project implementation; operations are physically distant from one another and the diversity of industries is low; most of the industrial operations are in close proximity, roughly over an area of about 16 km²; some of its main by-products are stored close to the local community/urban area, such as Alcoa's bauxite residue area, which can lead to greater community pressure for the industry to investigative innovative reuse approaches to reduce by-product footprint; a larger number of operations and a greater range of industry size, which on balance should be conductive to more synergy opportunities; it is located close to a major city, and thus a larger market for by-product reuse opportunities(Corder 2008).

There are still potential opportunities in Gladstone to make it a important example of a truly sustainable region. Industry could take control of these initiatives and manage the agenda that will drive the region down the sustainable development path. If industry chooses not to take on this role, conditions, such as a more challenging community, water scarcity and a carbon-constrained economy could force industry to work collectively to an agenda set by non-industry organizations, such as government, community groups or NGOs (Corder 2008, 6).

4.7 Rotterdam Europoort industrial area, the Netherlands

Rotterdam Europoort is also located by the sea, and, as in Aalborg, operates a district central heating system. Several mechanisms in relation to physical mattes are present in this industrial symbiotic case. Firms are physically interacting with nature and the environment when using alternative solutions in order to limit their industrial impacts. The Municipality maintains a physical link with firms when trying to foster the industrial symbiotic development by setting favourable regulations and adopting an open urban plan. Rotterdam Harbour has a system area: 10,000ha, implemented by Europoort/Botlek Interests industry association, where the actors are 80 industrial members; more than 30 chemical manufacturing firms and four refineries. The flows are heat and water and poorer availability of infrastructures (especially pipes) is thought to be one of the reasons for the limited development of the industrial symbiotic system at the beginning. Some synergies rely on very new technologies (complex heat distribution system through pipelines from firms to houses).

The Rotterdam Harbour and Industry Complex (HIC) (Baas 2008, 330–340) has been an environmental sanitation area in the period 1968–1998. The regional Environmental Protection Agency (EPA) and Water Authority regulate all firms in the area. Many, but not all, firms are involved in different covenants and are concerned with environmental performance targets, such as covenants on the reduction of hydrocarbons, the Chlorine Fluor Carbon reduction programme, the implementation of environmental management systems, and the four-year environmental management plan of a firm. »The Industrial Ecosystem (INES) project in the Rotterdam port industrial area started with the participation of 69 industrial firms in 1994« (Baas and Bones, 2004 1075–1085 quoted in Misra 2008, 147). The project was introduced by an industrial association Deltalinqs, active in the joint interests of industrial firms in the Europoort harbour and industry area near Rotterdam. Initially, the Deltalinqs' approach to environmental problems was very defensive. Later, a more constructive attitude was developed through the stimulation of environmental management in firms. In the period 1991–1994, some of the subsidies for the development and implementation of environmental management systems were used for the supervision of this implementation process. »The industrial association stimulated the acquisition of knowledge about environmental management and the feeling of responsibility of the firms through a communication structure involving meetings of environmental coordinators in six similar sectors of industry« (Baas and Huising 2008, 402). Following the national trend of self-regulation, Deltalinqs in 1989 started to develop an approach to promote environmental management systems in 70 member firms. During the period 1991–1994, it stimulated the firms' own responsibility through separate meeting groups for six branches of industry that meet quarterly. Facilitated by a consultant, firms exchanged information and experiences on the implementation of environmental management systems. In a coordinating group, experiences were exchanged among these groups. This structure was evaluated positively by the participating environmental coordinators of the firms (Misra 2008, 147). »Deltalinqs started to search for funds, which led to the start of the INES programme in 1994« (Baas and Huising 2008, 402). The INES Mainport 1999–2002 project took the possibility studies of the INES 1994–1997 programme, focused on the following issues: utility sharing, water, CO_2/energy, repose products/waste management, soil, and logistics (Baas and Boons 2007, 560–564). At the same time, a more strategic process was initiated. In 1998, the results from the INES programme were evaluated by the Board of Deltalinqs, which took time given that they met only twice every year (Baas 2011, 428–440; Almasi et al. 2011; Misra 2008). The project initiated a strategic decision-making platform, in which the societal actors from industry, government, EPAs, an environmental advocacy association, and a university were involved. The platform did meet occasionally, but was not actively approached to bring in new incentives.

The industrial symbiotic case in Rotterdam promotes public neutrality, which explains some of the absence of links among firms and the Municipality on this level. Thus, initiatives are mostly market-driven. Although, social considerations may come into play and produce a different outcome than if the decision would be only based on a rational ground.

In the Rotterdam Europoort case, environmental taxes have been applied concerning the use of certain rare materials. It had the effect of fostering the development of synergies aiming at reducing the consumption of such materials. From the Workshop that was edited by Industrial Symbiosis Institute in 2006, a person called Christensen Jørgen revealed that if tax-related incentives to reduce the use of rare materials are not given, the environmental regulations must not restrict the handling of waste (Almasi et al. 2011; Christensen 2010). The industrial symbiotic case reported the limitedly successful initiative of an external actor (INES) with the aim to find new synergies and »impose« them to firms. This promotes the importance of the firms »ownership of the synergy idea« (Almasi et al. 2011; Christensen 2010), meaning that a bottom-up management style is preferred to a top-down planning management, where synergy ideas come from firms themselves, acting in their own interests. In the industrial symbiotic system, we found a barrier that was the relatively long investment payback time of synergies involving new private firms (up to 5 years). Although the synergies were economically viable, the payback time was beyond private firms' business strategy scope, and therefore, hardly acceptable, especially since the liberalization of the country energy supply (Almasi et al. 2011; Christensen 2010).

The Rotterdam Municipal Port Management (RMPM) manages half of the land area. The decisions and evaluations concerning the various INES projects are made by a decision-making platform consisting of members of national and regional industry associations, plant managers, national and regional governmental organizations, an environmental advocacy organization, and academia. Europoort/Botlek Interests (EBB) industry association has an intermediary role between government and industries. INES was launched to stimulate the development of cleaner production approaches; to perform analyses of the activities, material, and energy streams; and to develop an information infrastructure to facilitate the functioning of the industrial ecosystem in the area.

The following achievements were made: a joint compressed air system was established (17 firms in 2004), the outsourcing of utilities as the core-business of service firms was made, and several industrial water systems were built. Regional efficiency was achieved in the industrial area during the INES Mainport programme. In addition, the following possibilities occurred: Intention to deliver 2,000 MW of heat (now emitted to the air) to 5,000 dwellings in 2006 (Baas and Korevaar 2010, 59–79). After it was determined that the establishment of a pipeline infrastructure for the whole area was economically not feasible, smaller scale projects were initiated in the INES Mainport project.

The Utilisation of Industrial Rest Warmth project involved eight partner projects in the Botlek and Pernis industry clusters. The estimated total investment was 83.6 million Euros. The Dutch National Project Office for CO_2 reduction plans was requested to provide a 30% subsidy in March 1998. A 27% subsidy was reserved in November 1998. A partnership of seven Deltalinqs firms tested the technical, operational and economic feasibility of the eight partners' projects during 1999 (Baas and Huisingh 2008, 399–421).

They decided to reject four projects, three for economic reasons and one on grounds of discontinuity of supply. At the beginning, despite the enormous waste heat surplus, nearly all managers of large plants had reasons to prefer their own facilities for economic (the costs of the required infrastructure) or strategic (the perceived loss of independence) reasons. That is why during the period 1997-2001, the waste heat supply project had to be downsized from a holistic regional approach to several small cluster projects. After this approach appeared to be economically failed, a feasibility study for heat delivery through a private »Heat Firm« was performed (ROM-Rijnmond 2003 quoted in Baas and Huising 2008, 406).

One of the drivers of the continuing effort to implement this theme was pressure from the Water Management Authority, who made it clear that they would no longer accept emission of heat into the surface water performed (Almasi et al. 2011).

5 Methodological framework

To answer the research questions and to define the hypothesis, a methodological framework has been developed. In this way, it is possible to analyse how industrial symbiotic networks develop. Using the detailed material of case studies, information can be gleaned about barriers, activities, social driving forces, and motivation approaches. Within these case studies, the theories of the development of industrial symbiotic networks will be used to define the specific changes in such systems. The theory functions therefore as a template to extract the significant data out of the case studies to detect the important facts that are significant for industrial symbiotic networks approaches.

The research departs from a fundamentally constructivist-interpretations paradigm. Preliminary literature overview shows that academic interest in industrial symbiosis was raised by examples of cooperating firms in practice in the second half of the 20th century. Nowadays a considerable number of case studies are available in literature from different countries. So, there is some contradiction on how industrial symbiosis is described in literature with its interpretation being the subject of critical discussions in published or issued researches (Baas and Boons 2007).

Most of the initial contributions focused on the engineering and technical feasibility of the exchanges, whereas social elements remained mostly unaddressed. We think that the missing gap in the literature is the branch of metaphysics (philosophy concerning the overall nature of what things are). It is concerned with identifying, in the most general terms, the kinds of things that exist in industrial symbiotic network, so industrial symbiotic foundations guide not only the selection of the research focus, but also the methodology adopted and the expected outcomes. Industrial symbiotic networks here are thus examined as a social structure that is simultaneously constructed and reflected upon by the involved authors and actors. This approach and the research understands their identification first as a singular phenomenon, and then as a sustainable instrument, part of the process of co-creation of meaning and social interaction (Džajić and Rončević 2017).

5.1 Empirical material and data analysis approach

The methodology developed is based on the concepts of industrial ecology and industrial symbiotic examples, but mainly from literature and web sources feedback from the industrial symbiotic projects that have been developed in last 25 years. The basis for model development is on industrial symbiotic studies from literature that can reflect the current state of industrial symbiotic networks as described in the literature.

The data collection is based on a mixed methods approach, conducted by the method of triangulation (qualitative-quantitative method) to ensure the validity of this study. This means that data are obtained from sources like literature reviews, empirical data, and from articles with interviews/practitioners who work with the facilitation of industrial symbiotic exchanges. A theoretical framework obtained from the literature on industrial symbiosis guides the data collection in all parts of the structure of this study. Literature review is an important data collection method and includes projects, legal texts, documents, reports, literature, scientific journals, studies, information from websites, and municipality reports. The collection of these data was taken from academic databases or/and libraries. Firstly, with the literature review, the basic methodology to gather data and knowledge is used for the foundation of this study, and these data will play an important role for the analysis and define the main characteristics to build a model (Booth, Colomb and Williams 2009; Schwarz 2018).

What is the relevance of choosing a quantitative method for studying the development of industrial symbiotic networks, and what exactly is the method of multi-attribute decision-making program? It is because this method allows:

- »Translate strategy into action«;
- The development of qualitative model to provide a quick overview over the development of industrial symbiotic networks across three different areas;
- The model on the (based on DEX) computational methodology and its implementation into the open source computer program (DEXi).
- According to the model, the evaluation of how well the prerequisites are fulfilled in some industrial area and the adoption of a deductive-based method where the reduction of reality (model) is confronted to reality itself;
- Easy understanding, not only for researchers in one of the areas discussed in this study, but it can be easily understood and interpreted by the public, initiators, and decision makers from different sectors; understanding and the perception of qualitative data of each other's businesses, business transparency issues, awareness of environmental pressure;
- The assessments of industrial symbiotic networking benefits (DEXi: A Program for Multi-Attribute Decision Making 2018).

The World Wide Web was used for literature research; science databases; published articles (some are published in *Journal of Cleaner Production*, Research-gate,

Journal of Environmental Management, Journal of Industry Ecology); online libraries; professional researches; dissertations; and industrial symbiotic projects already done and those which are underway. We searched for publications that were listed under keywords such as »Industrial Symbiosis«, »Industrial Symbiotic Networks«, »Industrial Symbiosis Networks«, »Eco-Industrial parks«, »Green Environment«, »Industrial Ecology«, or the combination of »Industrial Symbiotic Networks« as a topic. Seven real industrial symbiotic cases were chosen. From the resulting approximately 200 items, all entries that dealt with topics unrelated to material and energy flows among firms in regional industrial systems were removed. This procedure resulted in 116 publications that were analysed first for their conceptual background.

The publications related to the purpose of developing a theoretical structure for analysing and explaining the process through which industrial symbiotic networks are formed. The first category covers publications that offer conceptual structures or theories for understanding the empirical cases of industrial symbiotic networks.

Although it has been found that the scientific literature has partly addressed this gap and recognized the role of the social aspects, there is still little understanding of how social mechanisms work; how they affect the emergence and operation of industrial symbiotic networks; and, most importantly, there is a lack of comprehensive structures for the analysis of the soft elements of industrial symbiosis. Some publications offer conceptual structures or theories of a normative nature. These are designed and applied to prescribe how industrial symbiotic networks should develop, but they fail to properly analyse and explain specific empirical cases or allow the generalizability of results. The methodological structure outlined strives to capture the physical, organizational, and the most important social aspects that surround the process of emergence and development and the appropriateness of industrial symbiotic networks. The methodological part of this study focuses on the understanding of the social processes that govern material and energy exchanges in case studies. Therefore, due to the nature of the research questions, the methodology design takes a predominantly inductive approach in the empirical part, using a combination of qualitative and quantitative methods.

5.2 Decision problem development and implementation of the model

This study proposes design principles for tools that include support for the larger productive processes. The focus is on expert decision support tools for a specific case, in what we call creative problem solving in expert domains. By »expert domain«, we do not mean that everyone working in the domain is an expert, but rather that the domain itself is strongly associated with specialized knowledge (expertise). Domains that can be described as expert work share at least these six characteristics:

- The work requires specialized knowledge, typically obtained through formal education.
- The work is productive in the sense that it leads to the creation of artefacts.
- The work features creative problem solving as a central activity.
- The skills used in the work are predominantly non-recurrent (as opposed to the performance of well-practiced recurrent skills by, for example, pianists or tennis players).
- The work is intellectually challenging. It requires a broad range of intellectual skills and involves the gathering, organizing, analysing, and synthesizing of information.
- The product of the work is pragmatic rather than universal; the immediate goal is not to identify an abstract truth but to answer the needs of a specific situation (van Merriënboer 1997).

This study gains in stature by applying a feasibility development of the model to industrial symbiotic cases; and to compare the presumed fostering mechanisms of the new model to their degree of presence between the firms obtained in the focus area. This technique is used in order to reveal the already-in-place potentials and barriers to the development of industrial symbiotic networks. It allows an evaluation, according to the model, on how well the prerequisites are fulfilled in this area. It is achieved by adopting a deductive-based method where the reduction of reality (by means of modelling) is confronted by reality itself. This testing phase has been conducted to several levels using a software simulation tool.

In this study, the sphere of the physical (environmental) level (in the industrial symbiotic industrial areas) allowed us a collection of factual and visible information. It seemed to be the best way to detect physical relevant infrastructures in certain area. It confirmed (but not always) the presence of physical potentials and barriers. At an organizational and social level, the collection method is different. We as researchers constitute an external entity to the industrial sphere; it is hardly possible to measure (social or organizational) interactions between actors (as the model suggests) and structures without being an actor itself. The qualitative data about perception of each other's businesses, business transparency issues, awareness of environmental pressure, and industrial symbiotic networking benefits could be assessed (Džajić and Rončević 2017).

Firms need decision makers, managers who will give full scope to individual strength and responsibility and at the same time give a common direction of vision and effort, establish teamwork, and harmonize the goals of the individual with the common goods (Drucker 1955 quoted in Jereb, Rajkovič and Rajkovič 2005, 198–205; Milkovich and Boudreau 1997). This ensures that individual and corporate objectives are integrated and makes it possible for managers to control their own performance: »Self-control means stronger motivation: a desire to do the best rather than just enough to get by. It means higher performance goals and broader vision « (Armstrong and Baron 1998 quoted in Jereb, Rajkovič and Rajkovič 2005, 198–205).

The problem is »how to find a manager who will be able to translate business strategy into action«, but the decision to appoint such a manager requires the clear identification of the criteria (attributes) that distinguish successful from unsuccessful performance and use only predictive measures of success that are reliable and valid. Without a systematic approach that examines reliability and validity, no relationships can be demonstrated between selection criteria and selection predictors. Decisions thus remain subjective, of dubious value and open to challenge. In contrast, the stronger the relationship between predictors and criteria, the more accurate the employment decision and the easier it is to satisfy requirements that selection procedures must be objective, non-discriminatory, and result in the best candidate being selected (Beardwell and Holden 1997; Stone, 1998; quoted in Jereb, Rajkovič and Rajkovič 2005, 198–205). The choice of selection criteria (attributes) should be consistent with the organization's strategic direction and culture. Finally, as Cascio (quoted in Jereb, Rajkovič and Rajkovič 2005, 198–205) points out, »more accurate predictions result in greater cost savings (financial as well as social)«. The approach to support the selection of managers that is described in this study integrates multi-attribute modelling and expert systems. The solution of the selection problem is based on a multi-attribute hierarchical model and implemented with existing information technology. The basic principle of the multi-attribute decision making is a decomposition of the decision problem into smaller, less complex sub-problems. This makes the problem transparent and understandable; the decision becomes explainable and more effective. This method turned out to be very effective in different fields like banking, pharmacy, public administration, etc. (Klein and Methlie, 1995 quoted in Bohanec, Zupan and Rajkovič 2000, 191–205; Milner 2000 quoted in Jereb, Rajkovič and Rajkovič 2005, 198–205). »Such analysis of the situation is possible by the theoretical support given by work about social carriers of techniques for development« (Edquist and Edqvist 1979 313–331 quoted in Almasi et al. 2011, 30).

Then the use of the system in selecting a top manager is presented through: problem identification, project setup, modelling, options identification, options evaluation and analysis, decision making, and implementation.

Using the established system of descriptive city performance development indicators, we seek to enable qualitative decision making in a systematic way by using a multi-attribute model in complex situations with many factors and variables. In the first phase of the study, the relevant attribute/attributes were identified. Then we connected them into a hierarchical tree of attributes for building the decision model, where their qualitative scales were determined. In the first section, a rough version of the industrial symbiotic network model is defined upon 7 cases and 22 fostering mechanisms that are identified and explained, as well as the 3 dimensions they are embedded in. It is expected that these case descriptions would reveal similar fostering industrial symbiotic mechanisms, confirming the former model, but also other relevant mechanisms that would give a more comprehensive range of explanation. Due to the manner in which the information on individual industrial case exists, it was revealed that the most suitable method of construction of the multi-parameter model was the decision model.

5.3 Decision EXpert method and DEXi software tool

A used Decision EXpert method (DEX) in this study is a typical problem of the Multi-Attribute Decision-making methodology (MADM) (Multiple Criteria Decision Making, International Society on MCDM 2018). The MADM problems can generally be solved by several methods. Before we chose the DEX method, we took into account some other methods, such as the Analytic Hierarchy Process (AHP), Multi-Attribute Utility Theory (MAUT) for the quantification of subjective judgements in various fields of decision making, the Elimination and Choice Translating Reality (Elimination Et Choix Tradusiant la Realite) (ELECTRE), the Measuring Attractiveness by a Category-Based Evaluation Technique (MACBETH), the Preference Ranking Global Frequencies in Multicriterion Analysis (PRAGMA), the Preference Ranking Organization Method, etc.

The applications of different MADM methods or the combinations of these methods or even the adaptations of the fuzzy theory and the grey theory should be implemented, executed, and completed by the researchers in the different parts of the world to increase the number of the scientific studies and also the lessons learnt or experiences' statements based on the true scientific principles for the next generation (humanity) to help them to take more appropriate and more satisfactory decisions in all aspects.

The foundations of what eventually became DEXi were set up in Durham, UK, by Efstathiou and Rajkovič (1979, 326–333). Decision Expert-DEX was created when the method was implemented in a form of an expert system shell for decision-making (Bohanec and Rajkovič, 1990 145–157). This was an up-to-date implementation of the whole methodology. In the 1990s, DEX was used in a series of complex decision-making tasks in healthcare, projects management, industry, and sport (Bohanec and Rajkovič 1990, 145–157). In 2000, the implementation of DEXi started DEXi (Jereb, Bohanec and Rajkovič 2003) a somewhat stripped-down, but simple and user-friendly computer software tool, aimed primarily at education. This solidified DEX's entry into Slovenian secondary schools and universities (Krapež and Rajkovič 2003). Despite its simplicity, DEXi turned out to be extremely useful even for most difficult decision-making tasks.

The author choses the DEX method, because of the nature and the characteristics of the problem, appropriateness of the method for the modelling and solving the current problem, that could most easily be modelled by the decision of the qualitative attributes. Based on the real data, this problem could be assumed as a kind of complex real-world problems. The data and information could not be accepted as fully and wholly accurate and existing. The DEXi models were developed for these kinds of situations. Decision Support Methodology, as it is an approach that relies on interaction study with some main industrial symbiotic case studies and information, is continuously reinterpreted in an iterative process of collection and analysis of data. This methodology is a broad discipline concerned with supporting people – e.g. policy-makers – in making decisions. It is a part of decision sciences, together with normative and descriptive approaches to decision making (Clemen and Winkler 1999, 187–203; Jereb, Rajkovič and Rajkovič 2005, 198–205). Decision support encompasses several disciplines, including operations research, decision analysis, decision support systems, data warehousing, and group decision support (Jereb, Bohanec and Rajkovič 2003). The general approach originates in decision analysis, a discipline popularly known as applied decision theory. Decision analysis provides a structure for analysing decision problems (quoted in Jereb, Rajkovič and Rajkovič 2005, 198–205): »(i) structuring and breaking them down into more manageable parts; (ii) explicitly considering the possible alternatives, available information, involved uncertainties and relevant preferences; (iii) combining these to arrive at optimal or at least »sufficiently good« decision« (Bohanec 2008, 10).

5.3.1 Principles of DEX method and DEXi software tool

The decision maker had some specific properties, typical for this approach: (1) he deals with options or alternatives, i.e. entities or actions of a similar type in the sense that decision problems could be compared amongst themselves; (2) the goal is to select one option, or to evaluate or rank options in some preferential order; (3) the problem could be ranked into smaller, less complex sub-problems; (4) options could be described by basic features, i.e. vectors of values corresponding to the problem decomposition; (5) the evaluation of options could be represented by one or more mappings from basic features to one or more overall evaluations or classifications (DEXi: A Program for Multi-Attribute Decision Making 2018).

Hence, DEX methodology decomposes the complex decision problem at hand into smaller and easily understandable decision components, which are assembled into a hierarchical model. We can define types of attributes in DEX: one as *basic and the second as aggregated* (Damij et al. 2016, 6). »The former are the directly measurable attributes, also called input attributes that are used for describing the alternatives at hand« (Damij et al. 2016, 6). The latter are obtained by aggregating the basic and/or other aggregated attributes (Bohanec and Rajković 1990, 145–157; DEXi: A Program for Multi-Attribute Decision Making 2018; Damij et al. 2016).

In DEX, each of the n attributes is a discrete one i.e. the ith qualitative attribute QA_i, $i \in [1, n]$, can obtain values from a finite value set with N_i elements:

$$QA_i \in \{e_1^{(i)}, e_2^{(i)}, e_3^{(i)}, \dots e_n^{(i)}\}$$

The dependency among the attributes is defined by a utility function (Damij et al. 2016, 6–7). For each aggregated attribute y, the arguments of the corresponding utility function comprise of its immediate descendants from the hierarchical tree. Consequently, the utility function reads as (Damij et al. (2016, 6):

$$Y = u_y (QA_1, QA_2, QA_3, \dots QA_n)$$

Proper to the discrete nature of the attributes in DEX methodology, the utility function is usually represented as a qualitative decision table, as shown in Fig. 5.1. The modelling proceeded in consecutive sub-stages:

1. *Identification of the decision attributes*: The identification of decision parameters is a set of parameters that we want to pursue. Parameters are divided to the point that it is possible to assess within each alternative (option). Parameters are grouped collectively. For all the parameters, the measurements area is set (as in appendix, see column with Scales). Consequently, this involved option identification and option description (as in appendix, see the table Attribute tree). Options were collected and a database of options was created. We define relevant attributes using three spheres: environmental and technological, organizational and economy, and social. The result was an ordered list of all 22 (13 + 9) attributes (e.g. Tab. 5.1).

2. *Hierarchy of attributes* involved structuring, comparing, and aggregating the list of attributes using a »bottom-up« approach. Developing attributes' hierarchy was based on their interrelations and anticipated influence on the final proposed outcomes of the model. The result of this stage is the tree of attributes given in Tab. 5.2.

3. Definition of attributes' measurement: *Scales of attributes* and set of values that each attribute may obtain. The DEX method is a qualitative one and usually represents attributes whose values are represented by words, such as »impossible«, »incomplete«, or »possible«, or others (Tab. 5.2, column 2). Usually scales are ordered preferentially, i.e. from bad to good values.

4. *Capturing the decision knowledge*: Determining preferred parameters and then deciding how individual values of the parameters affect the final assessment. At this point, we are faced with the aggregation of partial evaluations for individual parameters in the final assessment. Parameters are combined due to content interconnections.

5. *Analysis and assessment of variants*: Evaluation variants of industrial symbiotic cases are nicely evident and easily comparable. It clearly shows all weaknesses, strengths, and good practices, making it easy to conclude. It is the degree of realization of the decision that must be monitored based on parameters and the decision-making skills (Bohanec and Rajkovič 1990, 145–157; DEXi: A Program for Multi-Attribute Decision Making 2018). Each option was evaluated and basic values were aggregated from the bottom towards the top of the hierarchy according to the decision rules. As a result, each option was assigned a qualitative evaluation value or, in the case that the option was described by imprecise or missing basic values, by a set of possible evaluation values. We used a technique »What-if« and »selective explanation«. The first assesses the effect of changing some basic attributes' value(s) to the partial or overall evaluation of a chosen option, and the second identifies the most

important advantages and disadvantages of options, which is important for the justification of the decision.

6. *Evaluation*: In Tab. 6.1, green coloured words mean higher value in the evaluation of industrial symbiotic cases, red coloured words stand for lower, and the black colour is medium evaluate attributes. Each score is a relative measure, which is not sufficient for the decision. It is necessary as an explanation of the differences between the various alternatives (Damij et al. 2016, 6–16).

The used decision support methodology in this study consequently relies on the model with some main industrial symbiotic real-life case studies and where information is continuously reinterpreted in an iterative process of collection and analysis data. The methodology is a comprehensive discipline concerned with supporting people in making decisions (Jereb, Rajkovič and Rajkovič 2005, 198–205). Theoretical structures are developed during the early stages of this study based on well-known examples of industrial symbiosis in the world. In accordance with industrial symbiotic principles, it appears that such an organization leads to forming an integrated natural/human system that helps minimize industrial activity impacts on the environment. Based on Tab. 5.1, we evaluated a new model »Development of industrial symbiotic network«. Tab. 5.1 shows the complexity of certain attributes that are outstanding, and a list of the main indicators to be applied in the comparative assessment of options; so, the indicators are predecessors of the sustainability evaluation criteria. The evaluation of attributes in Tab. 5.2 are constructed and determinate by author subjective consideration.

The hierarchical structure in DEX represents a tree. In the tree, attributes are structured so that there is only one path from each aggregate attribute to the root of the tree. The path contains the dependencies among attributes such that the higher-level attributes depend on their immediate descendants in the tree. This dependency is defined by a *utility function* represented in a tabular format such as the one presented in Fig. 5.1. The other evaluations of utility functions follow in the results chapter. Utility function is the component of multi-attribute model that defines the aggregation aspect of option evaluation. In Fig. 5.1, the utility function gives the aggregation rules for the attribute accessibility of material resources, based on the values of the input attributes availability of resources and the origin of those resources. For instance, one may read from the utility function in Fig. 5.1 that, if the origin of resources is »external« regardless of the availability of those resources (third row), then the accessibility of material resources that would be in line with the requirements for development of industrial symbiosis in the region is »low«. Also, if the availability of resources is none

Tab. 5.1: View of physical, organizational, and social aspects and measurements of coloured attributes. Source: Authors' own calculations 2018

Physical sphere

Environment and Technology fuses the availability and exchanges of material resources and the proximity of the physical plants. The longer distances between plants imply higher transportation costs. For example, if the exchange of material between plants involves transfers by pipelines, the capital costs of construction would increase in line with the distance of the plants. On the other hand, wheeled transports of materials incur less initial costs; however, the operating costs become more significant due to the increased noise, fuel usage, dust, and traffic. Due to the increased costs, the industrial symbiotic network development is modelled through the feasibility of: developed or the possibility for development of an accessibility of material resources in line with the industrial symbiotic requirements, the availability of resources, distances in the area where industrial symbiotic network should be developed, and proximity-vicinity of the businesses (Doménech 2010; Doménech and Davies 2011, 79–282; Ashton 2008; Almasi et al. 2011; Boshkoska, Džajić and Rončević 2018).

Accessibility of material resources is an attribute that describes the firms' access to industrial water, energy, and production materials (primary and secondary). Resource scarcity spurs resource efficiency and innovation. Resource efficiency refers to using the available and limited resources in a sustainable manner while minimizing the impact on the environment. Resource innovation means that firms focus on resource reduction and recovery (Corder 2008, 830–841; Ashton 2008; Almasi et al. 2011; Boshkoska, Džajić and Rončević 2018).

Availability of resources is an attribute that describes the availability of resources in the region such as water, energy, fossil fuels, steam, or production of raw materials (Cohen and Musnikow 2003, 2017). Due to the increased concentration of industries with high usage of the material resources for their productive processes, the pressure for availability of primary sources is increasing. This attribute may get values of »no« meaning that the region does not have any available resources; »partial« meaning that the region has limited resources; or »yes« meaning that regain has available resources and these are not fully used by the existing industries (Almasi et al. 2011; Cavallo et al. 2012; Boshkoska, Džajić and Rončević 2018).

In this model, the attribute *Availability of resources* may get values of no, partially, or yes. It is measured based on general characteristics of industrial symbiotic networks development such as:
1. High availability of water resources in the area.
2. Existing exchanges and reuse of different types of waste flows among firms.
3. The availability of critical resources such as energy or particular raw materials.
4. There are public and private benefits shown as »spontaneous co-location« (Chertow 2007) of businesses in industrial districts, to give rise to numerous public and private benefits including labour availability, access to capital, technological innovation, and infrastructure efficiency.

(*continued on next page*)

Tab. 5.1:   Continued

	5. Availability of close port/s that are used for acquiring resources. 6. Availability of industrial land. The value of the attribute Availability of resources is set as »yes« if the evaluated industrial symbiotic network satisfies 5–6 general characteristics described above; it has the value of »partially« if 3–4 general characteristics are satisfied; and if it satisfies 0–2 general characteristics above, then it has a »no« value.

Origin of resources refers to the way of connection of the available resources. We divided the origin of resources in internal, external, or mixed. Internal available resources include land, water, and all available natural resources amongst firms in the geographical connected industrial symbiosis. For example, when communities interact outside the political borders, they exchange and trade their internal resources to obtain external resources that are not available within their boundaries (political or natural) (Cohen and Musnikow 2003). For mixed available resources, we simply meant the combination of internal and external resources available in each presented industrial symbiotic case (Cavallo et al. 2012).

Proximity. According to Ayres and Ayres (2002), key industrial symbiotic network success factors are »collaboration and the synergistic possibilities offered by geographic proximity«. They differ from one industrial symbiotic network to another, and they have been largely discussed within the recent literature, often achieving slightly disagreeing results. Proximity is considered as an important factor in the process of development of an industrial symbiotic network. In the process of considering economical and technical feasibility among firms, they first carefully analyse their neighbours' input and output flows and try to find possible physical synergies. If they develop synergies with closely located industries, the infrastructure needed would be cheaper and their investment more profitable (Chertow 2007, 11–30; Ashton 2008; Almasi et al. 2011). For example, industrial symbiotic ecologists generally believe that close geographical proximity and trust are essential to the development of industrial symbiosis. To move industrial symbiotic research forward, this study suggests engagement with research in economic geography on the concept of proximity, which draws attention to the ways in which geographical, cognitive, institutional, social, and organizational distances between actors might affect innovation (Boshkoska, Džajić and Rončević 2018).

Distances are physical distances amongst firms. According to several industrial symbiotic cases, it seems that most synergies are taking place within a radius of 1–2 km distance between firms. A similar selection has been applied as a measurement scale to this attribute in the DEXi model {long, medium, short}.	The value of the attribute Distances is set as »short« if the evaluated industrial symbiotic network satisfies two general characteristics described above, it has the value of »medium« if we cannot confirm or deny any of two characteristics, and if it satisfies none of the general characteristics above, then it has an »long« value.

Tab. 5.1:   Continued

Having firms (Chertow 2007, 11–30) concentrated in a small area, having a short distance between them, can facilitate the feasibility of the industrial symbiotic process development. Long distances between firms can have an influence when making decisions (Cavallo et al. 2012).	In this model, the attribute *Distances* may get long, medium, or short values. When measuring distances, we considered: 1. The impacts of distance when making decisions among firms. 2. Metric distances between major cities and industrial centres in regional area. 3. Short distance referring to distances of up to 2 km, medium distances are distances of up to 20 km, and more than 20 km are considered long distance.
Installation/viability of heavy infrastructures refers to the vicinity of the main infrastructure suppliers to the industrial symbiotic area, including knowledge and technical expertise. Different types of infrastructures ensure a good circulation of flows between firms such as roads among firms, environmental infrastructures, water supply sewer and waste management units, pipelines, heat exchangers, cooling systems, etc. These components must work properly to have a successful industrial symbiotic system (Chertow and Lombardi 2005, 6535–6541; Ashton 2008). It is important to take into consideration also the vicinity of engineers and people having specific knowledge of close setting, because it can facilitate social networks in the participating firms.	In this model, the attribute *Installation/ viability of heavy infrastructures* may get values of impossible, incomplete, or possible. It is measured on the basis of general characteristics of industrial symbiotic networks development like: 1. The industrial symbiotic project does not have significant limitations of viability of transporting certain types of resources over long distances (such as steam or heat). 2. Sharing of access of heavy infrastructures (such as new waste exchange systems relying on pipeline or other heavy infrastructures; extensive roads, rails, and port infrastructures to allow for reliable transport) is in place (to reduce atmospheric emissions). 3. Geographical dispersion of firms enforces laws and regulatory codes. 4. Firms have an agreement to share the use of the installations. 5. The region has potential for renewable energy to a region's (country) natural capabilities. 6. There is a possibility for installation of heavy visible infrastructures. The value of the attribute Installation/viability of heavy infrastructures is set as »possible« if the evaluated industrial symbiotic network satisfies 5–6 general characteristics described above, it has the value of »incomplete« if 3–4 general characteristics are satisfied, and if it satisfies 0–2 general characteristics above, then it has an »impossible« value.

(continued on next page)

Tab. 5.1:   Continued

Organizational sphere

Organization and Economics refers to industrial symbiotic cooperation of firms with mutual benefit to cutting costs and improving the environment. This is an aggregated attribute that depends on _Economy and economics, Learning organization,_ and _Common industrial symbiotic communication platform_ amongst firms in the industrial symbiotic network. (Cohen and Musnikow 2003). Organizations may form different networking relations to complement each other by sharing knowledge and information, to have common businesses with selling or purchasing, and »to share common facilities or even cooperate in technological innovations« (Chertow 2007 11–30). Industrial symbiotic network could provide innovations through implementing activities to eliminate common environmental impacts and develop organizational collaboration and learning strategies (Jan 2002; Corder 2008; Uzzi 1996, 674–698; Golev, Artem and Corder 2014). At an organizational level, it is hardly possible to measure interactions between actors and structures (as the model indicates), because they are not an actor/structure. _Organization and Economics_ attribute is composed of mechanisms that are considered industrial symbiotic organizations in a metaphorical way; the theory suggests that such an organization is seen as an »organism adapting to its environment« (Chertow and Lombardi 2005, 6535–6541; Cavallo et al. 2012).

Economy and economics is one of the key motivations of a firm to join the industrial symbiotic networks and hence to cooperate in resolving existing environmental issues. The motivational drivers that led firms to establish environmental industrial symbiotic networks are basically the economic benefits such as cost savings, reducing costs by sharing investments in large joint infrastructural projects, ensuring availability of needed resources, such as water and energy, and implementing harsher legislative requirements (Barry 2008, Laybourn 2015). The economic benefits of industrial symbiotic networks are measured by determining the extent to which firms recycle by-products, for example, in the form of financial savings through the implementation of industrial symbiosis based on the reduction of various types of costs (cost of material, energy-related costs, waste-management costs, and costs of environmental legislation compliance). For example, when two firms cooperate in exchanging wastes, the economic benefits stemming from such cooperation can be quantified. The gross economic benefits stem from lower input purchase costs and lower waste disposal costs. There is a different cost sharing policy that a firm can use in exchanging wastes: firm X pays all the costs arising from industrial symbiosis, or firm Y pays all the costs arising from industrial symbiosis, or costs arising from industrial symbiosis are shared among firms.

Sustainable dynamics is an attribute that describes the possibilities to create economic growth, while advancing social and environmental objectives through an industrial symbiotic network. According to Schiller, Penn, and Basson (2014, 1–11), a sustainable community uses its resources to meet current needs while ensuring that adequate resources are available for future generations.	In this model, the attribute _Sustainable dynamics_ may get values of weak, partial, or strong. It is measured on the basis of general characteristics of industrial symbiotic networks development such as: 1. A firm's objectives are directed towards sustainable developments for the industrial symbiosis to have success. 2. Sustainable dynamism in the firm includes integrated policy, planning, and social learning processes.

Tab. 5.1: Continued

It involves all its citizens in an integrated, long-term planning process to protect the environment, expand economic opportunities, and meet social needs.	3. Sustainable dynamics and objectives are successfully integrated into the policy programs of government. 4. Practical implementation of sustainable principles achieves sustainable objectives – decision-making processes. Environmental, economic, and social objectives have oriented a firms' efforts in the reduction of main environmental impacts on a continuous process and have contributed to the reinforcement of the firms' commitments (Doménech 2010). The value of the attribute Sustainable dynamics is set as »strong« if the evaluated industrial symbiotic network satisfies 4–5 general characteristics described above, it has the value of »partial« if 3–4 general characteristics are satisfied, and if it satisfies 0–2 general characteristics above, then it has a »weak« value.
Reduction of raw material and cost saving means that firms on an individual level have outcomes in terms of the reduction of raw material, the reuse and recycling of products that lead to cost savings and have a positive impact on the environment. When being part of an industrial symbiotic system, the firm tends to reorganize their production process so that the wastes or by-products of one firm become the raw materials for another. One example of this was described by Kalundborg, where »60 million dollars of investment in Eco-industrial network infrastructure generated 120 million dollars in cost savings over five years. The so-called »win-win« situation is transferring firm's outputs to others thereby avoiding landfills disposal costs. This exchange can be remunerated and can contribute to a rise in the firms' incomes« (CRESSI Publications Saïd Business School No. 5. 2015; Social Innovation, Individuals and Societies: An Empirical Investigation of Multi-layered Effects 2017; Jacobsen 2006, 239–255 6; Jacobsen and Anderberg 2005, 31–335).	In this model, the attribute Reduction of raw material and cost saving may get values of weak, partial, or strong. It is measured on the bases of general characteristics of industrial symbiotic networks development such as: 1. Reuse of raw material can significantly reduce the industry vulnerability with regard to raw material scarcities (for example, water, greenhouse gases). 2. Initiatives are put in place to resolve the reduction of raw material and cost saving situation. 3. Providing real value of the money to the firm in some way, such as through cost savings or increased revenue. 4. Implementation of collaborative technologies for the reduction, reuse, and recycling of raw material. The value of the attribute Reduction of raw material and cost saving is set as »strong« if the evaluated industrial symbiotic network satisfies 3–4 general characteristics described above, it has the value of »partial« if 1–2 general characteristics are satisfied, and if it satisfies none of the general characteristics above, then it has a »weak« value.

(continued on next page)

Tab. 5.1:   Continued

Learning organization is explained as a concept where one has an ease of making and breaking connections as conditions change. In other words, it enables the inclusion of innovative ideas by employees into the decision-making process. Firms' management is usually aware of the usefulness of networking as a way of obtaining knowledge and information sharing. New approaches and ideas for firms' management can be drawn from gaining new contacts from other firms. Networks could be created as a knowledge pool from which the existing and new firms could benefit (Corder 2008; Steward 2008; Almasi et al. 2011).

Participative management describes an approach that allows employees to participate with suggestions in the development of the firm. To recognize the employers' ideas, firms organize workshops and brainstorming sessions with different technicians and engineers to collect new feasible ideas concerning potential synergies (Chertow 2007; Ashton 2008). Employees who are not members of the management teams are included in the decision-making process by allowing them to produce new ideas in the process of product reengineering.

Innovative ideas are attributes that describe cases when employees are allowed to produce new, innovative ideas in the process of product reengineering. Namely many firms are aware of the potential of their employees and customers' knowledge, experience, and ideas that may lead to improved industrial symbiotic networks (Boshkoska, Džajić and Rončević 2018).

In this model, the attribute *Innovative ideas* may get values of weak, partial, or strong. It is measured on the basis of general characteristics of industrial symbiotic networks development such as:
1. Accumulated know-how developed through the long experiences and relationships of employees.
2. Sharing of information and know-how among employees.
3. Collaboration and dialogue in firms resulted in new ideas.
4. Encouraging creativity and innovation of employees.
5. Common cooperation is associated with the development of industrial symbiotic networks of cooperative relationships between people, acting essentially as individuals, exchanging ideas, knowledge, and advice in firms.
6. Firms consider promoting their employees' efforts to improve, make progress, and evolve in the environmental performance of the organization they represent.

Tab. 5.1:   Continued

	The value of the attribute Innovative ideas is set as »strong« if the evaluated industrial symbiotic network satisfies 5–6 general characteristics described above, it has the value of »partial« if 3–4 general characteristics are satisfied, and if it satisfies 0–2 general characteristics above, then it has a »weak« value.
Common industrial symbiotic platform: These platforms, also called pools or communications centres, aim to provide communication as well as a cooperation platform for the firms in the industrial symbiotic networks. These platforms offer to the firms extra professional activities and better social links. They were developed to reach out beyond the existing industrial symbiotic networks and communication between managers, by measuring the in/out degree (ties claimed by others about the actor and ties claimed by the actor about others, respectively). From literature, it has been found that some authors use these platforms to analyse the average degree (number of ties per firm), density (ratio of actual ties to all possible ties), and average constraint (measures how constrained each node is by its neighbours) (Jereb, Bohanec and Rajković 2003a, 2003b; Boshkoska, Džajić and Rončević 2018) of the industrial symbiotic network.	In this model, the attribute Common industrial symbiotic platform may get values no, partial, or yes. It is measured based on general characteristics of industrial symbiotic networks development such as: 1. Existing institutional platforms and linkages, communication and trust, and coordination. 2. Comprehensible databases: Both existing and potential businesses can easily identify possible synergistic connections. 3. The available inventory and assessment of potential symbiosis connections (Rehn 2013). The value of the attribute Common industrial symbiotic platform is set as »yes« if the evaluated industrial symbiotic network satisfies 3 general characteristics described above, it has the value of »partial« if 1–2 general characteristics are satisfied, and if it satisfies none of the general characteristics above, then it has a »no« value.

Social spheres

Social forces consider different aspects from the social theory including *Institutions, Cognitive frames, and Networks* that »are not the only devices to resolve coordination problems in market fields however play a major role in defining the industrial symbiotic networks« (Beckert 2009, 245–269).	To promote industrial symbiotic networks, it is important for institutions or external agencies (e.g. industrial symbiotic institutes/eco-town centres/universities) to link with the social platform of industry, so that they can establish a good relationship and trust with one another for easier information exchange and the development of synergies. Also, holding periodical actors' discussions, public dialogue with industries promotes industrial symbiosis and facilitates match making.

(*continued on next page*)

Tab. 5.1: Continued

»While some firms have a great benefit from the existing network composition, institutional rules, and cognitive mind-sets in the market field, others are disadvantaged. Hence, the industrial symbiotic actors (decision makers) are engaged in an ongoing struggle to change or to defend the social forces operating in the field. Social aspects in industrial symbiosis also involve groups, industries to exchange by products and sharing utility, and interactions among actors« (Beckert 2010, 611–612; Fligstein 2002). Social platforms (e.g. environments/social clubs/associations) that engage different industries to come together to share the common understanding and build networks between firms are important.	This will help to cultivate their interest and willingness to actively participate in potential synergies. Building a strong social cohesion among industries takes time. The model provides an insight of industrial symbiotic networking that presents social interactions as social forces; institutions (laws, central concern for law, the formal mechanism for political rule-making and enforcement…); networks (social structure made up of a set of social actors, such as individuals or firms and a set of the dyadic ties between these actors); and cognitive frames (social interaction, making technologies, and strategically selective opportunities for reflection and learning) (Chertow 2000, 2007).
Institutions are often defined by rules and shared understandings that structure market exchange, encompass formal parameters such as labour laws; subsidies; intellectual property rights; and industry standards as well as informal norms, routines, ethics, and conventions that shape the practice of market actors (Fligstein 2002). Actors that do not trade but regulate markets implement some of these institutions. Other institutions are established by sellers, buyers, workers, cultural intermediaries, or consumers who populate a field and who have settled on arrangements to produce, distribute, evaluate, and consume certain products, either because of conscious agreements or because of what field participants perceive to be »normal« (Corder 2006).	Institutions are considered as social forces and are primarily understood as cultural scripts providing orientation for actors under conditions of uncertainty (Chertow and Lombardi 2005). The alternative concept of institutions, conceived as state-devised formal rules that regulate and constrain the behaviour of firms (Black Sea Industrial Symbiosis Platform 2017), forms the conceptual background of historical institutionalism and comparative political economy but finds limited attention in the new economic sociology (Bohane et al. 2000).
Urban planning is a sub-attribute of _Institution_ and describes the conditions of granting planning permissions and the organization of the overall land use, protection and use of the environment, public welfare, as well as providing a design of the urban environment.	In this model, the attribute _urban planning_ may get low, moderate, or satisfying values. It is measured on the basis of general characteristics of industrial symbiotic networks development such as:

Tab. 5.1:   Continued

The *Urban planning* is created when new industrial building areas are constructed and, in terms of industrial symbiosis, tells how and for what a specific area is used for and also where pipes and electricity should be placed. Urban planning can be altered by the municipality after conducting several legal steps. The openness and easiness to modify an urban plan to allow the building of new industrial symbiotic network infrastructures heavily depends on the area and the preferences of the decision makers on a municipal level (Boshkoska, Džajić and Rončević 2018; Jereb, Bohanec and Rajkovič. 2003a, 2003b). »For example, when planning a new waste exchange systems or heavy infrastructures, changes in the urban plan need to be approved and followed by municipally permission and local population« (Shi, Chertow and Yuyang 2010, 191–199; Almasi et al. 2011).	1. Existence of a controlled rural urban migration, urbanization with proper planning, and rapid industrialization with good infrastructure. 2. Existence of an on-site waste management plans. 3. Proper planning for required material quantities, on time passing of information on types and sizes of materials and components to be used. 4. Good supervision of the urban planning (Letcher and Vallero 2011). The value of the attribute Urban planning is set as »satisfying« if the evaluated industrial symbiotic network satisfies 3–4 general characteristics described above, it has the value of »moderate« if 2 general characteristics are satisfied, and if it satisfies 0–1 general characteristics above, then it has a »low« value.
Regulation and waste management describes the legislative matters setting limitations in terms of possibilities and quantities with regards to flows exchanged between industrial symbiotic firms (Schiller, Penn and Basson 2014, 1–11). Enforcing waste disposal regulations can trigger networking and synergies among firms, so that they are able to cope with all imposed limitations. The process of obtaining permission is time consuming (Cohen and Musnikow 2017). For example, the procedure for obtaining permission can be described in four consecutive steps. First, a draft proposal for a new local area change is submitted. Afterwards, the proposal has to be accepted by the City Council (or equivalent body). Next the new urban plan including the proposal is published in order to obtain the public opinion.	In this model, the attribute Regulation and waste management may get values of fully enforced, partially enforced, or weakly enforced. It is measured on the basis of general characteristics of industrial symbiotic networks development such as: 1. The existence a governmental action and commitment for reform towards green growth. 2. Emergence of a variety of eco-innovations because of regulatory and market-based instruments. 3. Enforcement of environmental taxes and regulations on harmful substances. 4. Literature evidence that government regulations and liability concerns are one of the reasons for the action in developing synergy initiatives.

(continued on next page)

Tab. 5.1:   Continued

Finally, the municipality administration agrees on the plan changes and publishes it. Environment laws and regulations can be introduced to encourage industry to adopt environmental technology and form symbiotic linkages. Industrial symbiosis becomes more economically feasible when it is enforced through legislation. Establishing relevant taxes, fees, and levies are a motivation tool for development of an industrial symbiosis. The environment taxes on certain raw resources foster the development of synergies to consume fewer raw materials (Chertow and Lombardi 2005).	The value of the attribute Regulation and waste management is set as »fully enforced« if the evaluated industrial symbiotic network satisfies 4–5 general characteristics described above, it has the value of »partially enforced« if 2–3 general characteristics are satisfied, and if it satisfies 0–1 general characteristics above, then it has a »weakly enforced« value.
Cognitive frames: According to Beckert (2009, 254–269), »cognitive frames are one of the three irreducible social forces that exercise their influence through constituting the perception and legitimation of institutional forms and network structures«. »In the social sciences, cognitive framing comprises of a set of concepts and theoretical perspectives on how individuals, groups and societies, organize, perceive and communicate about reality«. Moreover, they are shaped specifically within social networks (Levin and Gaeth 1988, 374–377; Boshkoska, Rončević and Džajić. 2018; Almasi et al. 2011; Ashton 2008). »Cognitive framing involves social construction of a social phenomenon, through mass media sources, political or social movements, political leaders, or other actors and organizations. In social theory, framing is a schema of interpretation-a collection of anecdotes and stereotypes-that individuals rely on to understand and respond to events« (Beckert 2009, 254–269). In other words, people build a series of mental »filters« through biological and cultural influences. They then use these filters to make sense of the world.	This attribute may get values of weak, medium, or strong and has been measured according to some general characteristics of industrial symbiotic networks that were elicited from the literature including: 1. The changes of environment may lead to the marginalization of the firms in cases when firms are not adaptable to the changes. 2. Some insights are present on the interrelation and dynamics amongst cognitive frames and other social forces. 3. Highly innovative solutions that alter cognitive frames might outweigh long-established institutional incentives like tax exemptions. 4. There are present and implemented sanctions that are aimed to foster the integration of people into the employment force in industrial symbiosis. 5. »Shared meaning structures might be incorporated into the notion of institution and therefore cognitive frames are not explicitly distinguished« (Beckert 2010, 605–527).

Tab. 5.1:   Continued

The choices they make are influenced by their creation of a frame (Boshkoska, Džajić and Rončević 2018; Social constructionism 2018, Boshkoska, Džajić and Rončević 2018; CRESSI Publications Saïd Business School No. 5. 2015).	We set the value of the attribute Cognitive frame as »strong«, if the industrial symbiotic network satisfies 4–5 general characteristics above. If industrial symbiotic network satisfies 2–3 general characteristics above, then it has a »medium« value, and if it satisfies the 0–1 general characteristics above, then the value is »weak«.
Networks are social relations (personal or organizational) between market actors (producers, consumers, suppliers, regulators, workers) and express the mental maps and the structure of the existing social relations (Beckert 2010, 605–527). The objectivity of networks is not constituted by the position of links and the structure of their connections as such, but by the dominant interpretations through which actors perceive the network structure. Networks generate trust, facilitate the exchange of information, or reduce risks. The opposite can be a comparison with ties that »are founded solely on cost calculations. Network relations and based on the expectation of reciprocity« (Granovetter 1985, 481–510). In this model, the attribute *Networks* may get values weak, medium, or strong (Boshkoska, Rončević and Džajić 2018).	It is measured on the basis of general characteristics of industrial symbiotic networks such as: 1. The number of different public sectors or actors that can support industrial symbiotic developments in different ways. 2. The network is considered strong if the public sector can set ambitious local policies demanding improved waste management and/or reduced emissions, creating the context for symbiotic exchanges. 3. The public sector can influence the highly effective relationships and information brokers, by creating vital conditions for communication, familiarity, and trust among regional actors in industrial symbiosis. 4. The planning in industrial symbiosis, and procurement functions, can also be adapted to create more fertile contexts for the development of industrial symbiotic networks. 5. Networks can support or hinder symbiotic relationships relevant to industrial symbiotic networks, with a focus on the influence of the policy structure like the nature of inter-firm business models and governance mechanisms, and the role of public–private partnerships (Granovetter 1985, 481–510).

(continued on next page)

Tab. 5.1:   Continued

Industrial symbiotic networks are »embedded« (Baas and Huisingh 2008, 399–421; Kalundborg Symbiose 2017) in social systems and, as such, decision-making processes are shaped by social relations (regulation systems, trust, beliefs, and knowledge are crucially influencing the direction and management of physical exchanges) (Jacobsen 2006, 239–255; Jacobsen and Anderberg 2005, 313–335; Boshkoska, Rončević and Džajić 2018). The value of the attribute *Networks* is set as »strong« if the evaluated industrial symbiotic network satisfies 5–6 general characteristics described above, it has the value of »medium« if 3–4 general characteristics are satisfied, and if it satisfies 0–2 general characteristics above, then it has a »weak« value (Boshkoska, Rončević and Džajić 2018).

(regarded as »no« in the model), then regardless of the value of the attribute origin of resources, the possibility of development of a suitable accessibility of material resource is »low« (DEXi: A Program for Multi-Attribute Decision Making 2018). To have at least moderate accessibility of material resource, the actual availability of resources should be evaluated at least as »partial« and the origin of resources should be at least evaluated as »mixed« (Boshkoska, Rončević and Džajić 2018).

An example of such a decision rule is given in the first row:

IF *Availability of resource is »yes«* **AND** *Origin of resources is »mixed«* **THEN** *Accessibility of material resource = »moderate«.*

The percentages in the second row are weights of the attributes. In the software tool DEXi, the utility function values are either entered by the user or not entered. Entered values are never changed by DEXi during the editing in the software tool. This is, however, not the case with non-entered values, which are handled by DEXi with the purpose to aid and simplify the function editing process and maintain the consistency of function definitions. Non-entered

Tab. 5.2: DEX model tree, scale attributes, and short description of attributes. Source: Authors' own calculations in DEXi 2018

Attribute	Scale	Description
Industrial Symbiotic Network Development	Partially developed; mostly developed; fully developed	Current Industrial Symbiotic Network Development
Environment and Technology	Weak; moderate; strong	Facets, location, diversity of industries, presence of river, sea
Accessibility of material resources	Low; moderate; satisfying	Scarcity of resources
Availability of resources	No; partially; yes	Industrial water, energy, production materials (primary, secondary)
Origin of resources	External; mixed; internal	How the available resources are connected: internal, external, mixed
Proximity	Low; moderate; satisfying	Physical distance among industrial installations
Distances	Long; medium; short	Distances among firms
Installation/viability of heavy infrastructures	Impossible; incomplete; possible	Vicinity of the main infrastructure suppliers to the industrial symbiotic area
Organization and Economics	Low; moderate; satisfying	Economic benefits, profitability
Economy and economics	Low; medium; high	Yielding the benefits of industrial symbiosis
Sustainable dynamics	Weak; partial; strong	Create economic growth, while advancing soc. and environ. objectives
Reduction of raw mat. and cost saving	Weak; partial; strong	Reduction, reuse and recycling; Income generating synergies
Learning organization	Low; moderate; satisfying	An approach includes new ideas in the decision-making process
Participative management	Weak; partial; strong	Involvement of employees into decision-making processes
Innovative ideas	Weak; partial; strong	Employees generate innovative ideas
Common industrial symbiotic platform	No; partial; yes	Existence of a joint and comprehensive IS platform
Social Forces	Low; moderate; satisfying	Social factors shaping the topography of ISNs
Institutions	Weak; partial; strong	Rules, ritualized behaviours
Urban planning	Low; moderate; satisfying	Local and/or regional spatial planning
Regulations for waste management	Weakly; enforced	Implementation of the waste management regulations
Cognitive frames	Weak; partial; strong	Green values, public awareness, and green procurement
Networks	Weak; partial; strong	Social trust and ease of cooperation

Availability of resources	Origin of resources	Accessibility of material resources
43%	57%	
1 no	*	low
2 <=partially	<=mixed	low
3 *	external	low
4 yes	mixed	moderate
5 >=partially	internal	satisfying

Fig. 5.1: Utility functions of the attribute Accessibility of material resource. Source: Authors' own calculations in DEXi 2018

values are shown in normal typeface and, by default, they are recalculated whenever the table changes. The calculation is based on already entered values and other available information (particularly weights). The difference in DEX is that attributes are symbolic and utility functions are defined by decision rules. Figure 5.1 consequently illustrates the basic approach. It shows the attribute accessibility of material resource utility function. The weights of attribute availability of resources and attribute origin of resources are almost identical, 43% and 57%, respectively. These are called local normalized weights (DEXi: A Program for Multi-Attribute Decision Making 2018; Boshkoska, Rončević and Džajić 2018).

Based on Tab. 5.1, a multi-parameter hierarchical decision model with three branches is produced in Tab. 5.2 that covers three main areas: the physical sphere, organizational sphere, and social sphere. This initial stage is entered as a result of becoming aware that a decision-making problem has occurred that is sufficiently difficult, important, or both to require a serious, careful, and systematic approach (Bohanec 2003; Jereb, Bohanec and Rajkovič, 2003a).

Bellow in Fig. 5.2 is presented DEXiTree, a small and publicly available software tool that provides four state-of-the-art tree-drawing algorithms. The development of DEXiTree has been directly motivated by DEXi, by its current inability to make nice drawings of multi-attribute models (Kt.ijs 2018). DEXiTree provides a rich set of parameters for an interactive design of the visual appearance of trees and their components: nodes, arcs, and text boxes. Namely, in DEXi, terminal nodes represent input attributes of a multi-attribute model, so it makes sense that they are grouped together and shown at a same level.

Also, in Fig. 5.2, the DEXiTree is oriented from left (root) to right (terminal) nodes. In this case, each terminal node occupies one »line« in the drawing, producing a nice and highly readable layout (Kt.ijs. 2018).

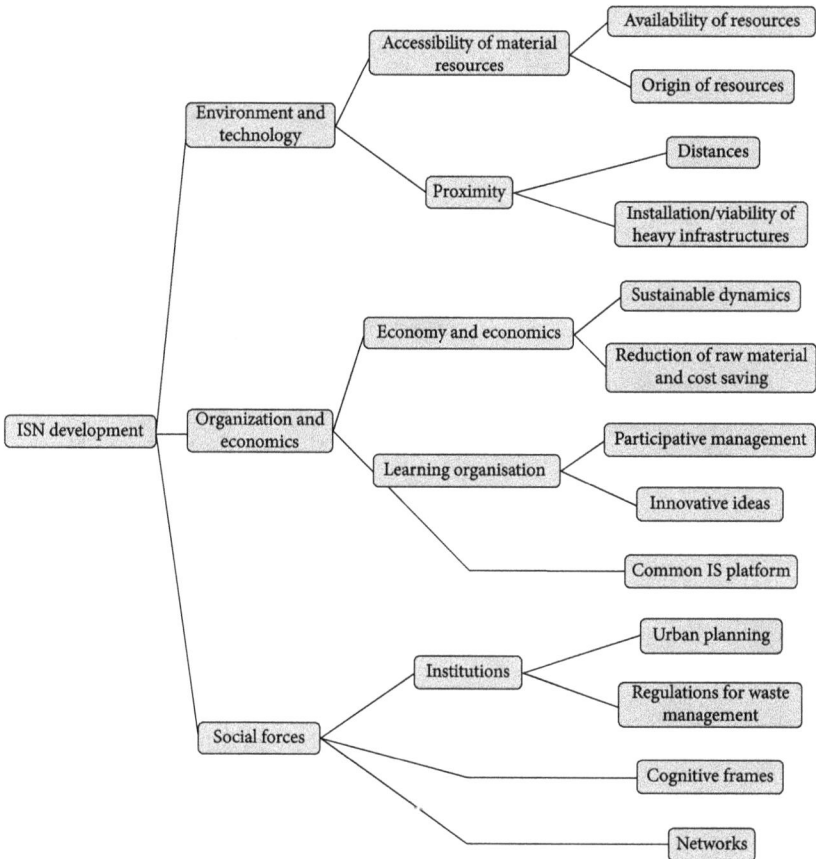

Fig. 5.2: DEXiTree report. Source: Authors' own calculations in DEXiTree 2018

6 Modelling results and evaluations of case studies

The (new) model is based on seven developments of industrial symbiotic networks. All indicators are selected from the studies of those industrial symbiotic cases. For each of the cases, we have presented how the values of the attributes are obtained. It is very difficult to measure each attribute quantitatively, hence we used a qualitative approach to measure the attributes. Even more, we have defined several qualitative indicators for each of the attributes and counted how many of them are fulfilled. The indicators are drawn from the literature on the seven industrial symbiotic cases, so that we define the indicators as exhaustively as possible, and at the same time to avoid focusing only on one industrial symbiotic network case.

The empirical framework is designed and the qualitative data indicate the research questions outlined. The ontological premises and the epistemological social foundations of the research are addressed. The main limitations of the methodological approach adopted are carefully examined. Basic concepts required for modelling sociological phenomena are used, and the data are collected from secondary empirical case studies. The process of selection (based on physical, organizational, and social conditions) of case studies, the criteria used, and the decisions of appropriateness of industrial symbiotic networks ensure clear identification of the criteria (attributes). These quantitative data can help distinguish successful from unsuccessful performances of industrial symbiotic networks and use only reliable and valid predictive measures of success. For better explanations, and to ensuring the validity of results, we have described these indicators for all input attributes in Tab. 5.1. Industrial symbiosis is a complex phenomenon; its development is difficult to explain through a set of indicators and therefore, we tried to additionally explain the set of indicators that are used to measure the attributes in the model. This is the first DEX model, which may be further developed; however, we wanted it to be as simple as possible in this phase, so that it can be used as a general model that incorporates the most important views of an industrial symbiotic network (organizational, economic, and social views). We have improved the description of all input attributes; we have defined all indicators for the input attributes, and provided and explained all utility functions of the aggregated attributes (Boshkoska, Rončević and Džajić 2018).

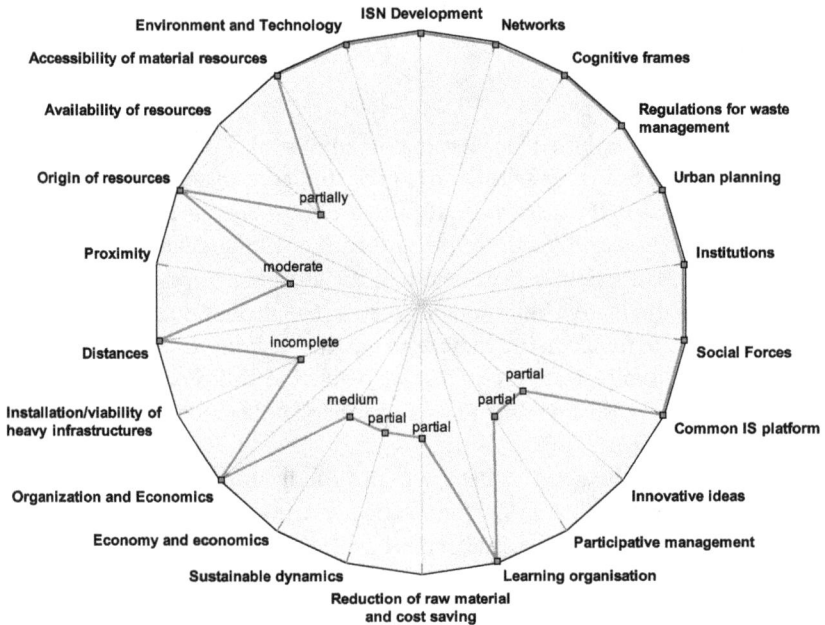

Fig. 6.1: Polar charts of the selected attributes for industrial symbiotic network in Kalundborg. Source: Authors' own calculations in DEXi 2018

According to the DEX model, the industrial symbiotic network in **Kalundborg** is evaluated as »fully developed«. The polar map of the selected attributes for Kalundborg is given in Fig. 6.1. Attributes include: environment and technology, institutions, cognitive frames, and networks are evaluated as »strong«; accessibility of material resources, organization and economics, learning organization, social forces, urban planning is rated as »satisfying«; the availability of resources is »partially« evaluated; the origin of resources is »internal«; proximity is »moderate«; distances are »short«; installation/viability of heavy infrastructures are »incomplete«; economy and economics are evaluated as »medium«; sustainability dynamics, reduction of raw materials and cost savings, participative management, and innovative ideas are »partial«; there is a common industrial symbiotic platform; and regulations for waste management are »fully enforced«.

According to the DEX model, **Aalborg** has a »mostly developed« industrial symbiotic network due to the incompletion of the attribute installation/viability of heavy infrastructure, the partial reduction of raw material and cost savings,

Aalborg

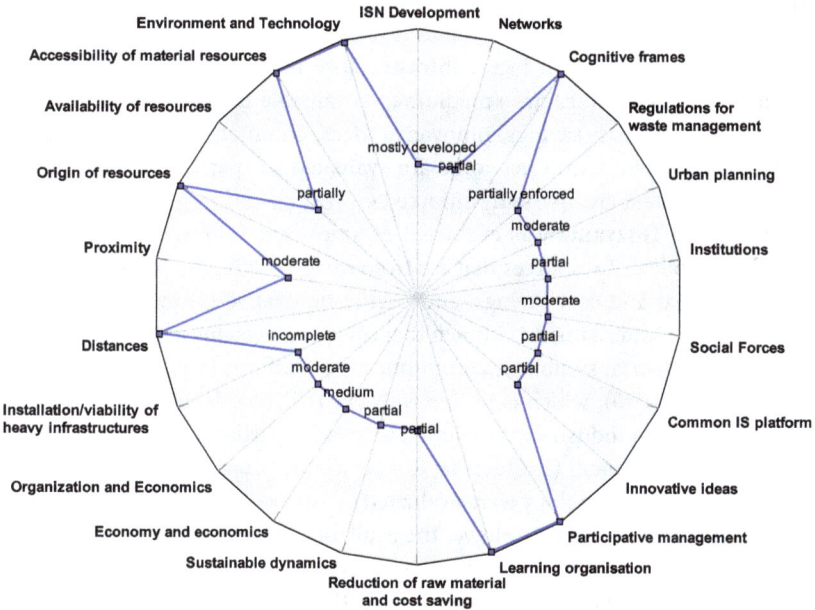

Fig. 6.2: Polar chart of the selected attributes for industrial symbiotic network in Aalborg. Source: Authors' own calculations in DEXi 2018

and the partial common industrial symbiotic platforms. These results are in line with the literature, which states that because Aalborg East does not face too many problems with natural resource scarcity, firms are not significantly motivated to decrease their resource consumption (Boshkoska, Rončević and Džajić 2018). Furthermore, there was no data about input/output or massive material flows in this area, so firms tended to hold a sceptical or curious attitude with regards to what extent other firms would share information and knowledge about within the area and protect their business confidentiality (Chertow and Lombardi 2005, 6535). Firms in this area had few extra-professional activities and relatively weak social links, which limited the space for developing deeper collaboration and mutual trust (Barry 2008). The polar map of the selected attributes for Aalborg is given in Fig. 6.2.

Attributes: environment and technology, participative management, and cognitive frames are evaluated as »strong«; accessibility of material resources and learning organization are »satisfying«; availability of resources is »partially« evaluated; the origin of resources is »internal«; proximity, organization and economics, social forces, and urban planning are »moderate«; distances are »short«; the installation/viability of heavy infrastructure is evaluated as »incomplete«; economy and economics are »medium«; sustainable dynamics, reductions of raw material and cost savings, innovative ideas, common industrial symbiotic platform, institutions, and networks are evaluated as »partial«; regulations for waste management are »partially enforced«.

The case of **Guayama** was evaluated as »mostly developed« because of the partial availability of resources that are from mixed origin, the partial reduction of raw material and cost savings, and partial innovative ideas. In the cases of Guayama and Rotterdam, it is known that there is a need for involvement of actors in the industrial symbiotic communication platform (Baas 2008, 330–340; Baas 2011, 428–440), which is in line with the DEXi model evaluations for the attribute common industrial symbiotic platform as partial suggesting the possibilities for improvement (Boshkoska, Rončević and Džajić 2018). The attributes environment and technology were evaluated as »weak«; the accessibility of material resources is evaluated as »low«; the availability of resources is evaluated as »partially«; the origin of resources are »mixed«; proximity and social forces are »moderate«; distances are something in the middle – »moderate« was the evaluation; installation/viability of heavy infrastructure is evaluated as »incomplete«; organization and economics, learning organization, and urban planning are »satisfying«; economy and economics are »high«; sustainable dynamics and networks are »strong«; the reduction of raw material and cost savings, participative management, common industrial symbiotic platform, institutions, and cognitive frames are »partial«; the regulations for waste managements are »partially enforced« (Boshkoska, Rončević and Džajić 2018). The polar map of selected attributes for Guayama is given in Fig. 6.3.

Industrial symbiotic network in **Barceloneta** is evaluated as »fully developed«; however, due to the medium distances the attribute proximity is evaluated as moderate because according to Velenturf and Jensen (2016, 700–709), industrial ecologists commonly believe that trust and geographical proximity/vicinity are essential to the development of industrial symbiosis and consequently industrial symbiotic networks. Close geographical proximity and confidence amongst actors and firms are essential to the development of industrial symbiosis, according to Velenturf and Jensen (2016, 700–709). Attributes environment and technology, sustainable dynamics, innovative ideas, institutions, and networks« are evaluated as »strong«; the accessibility of material resources, organization and economics, learning organization, social forces, and urban planning are

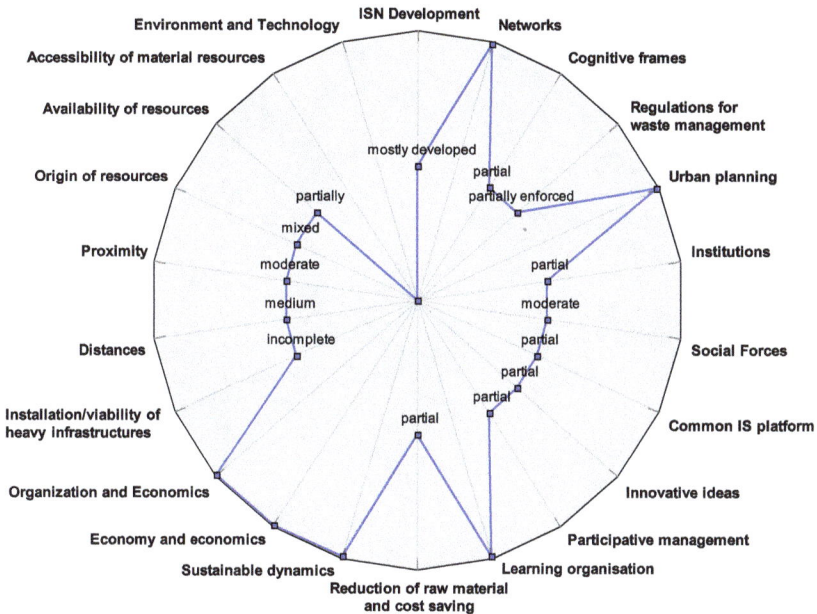

Fig. 6.3: Polar chart of the selected attributes for industrial symbiotic network in Guayama. Source: Authors' own calculations in DEXi 2018

»satisfying«; there is a lot of availability of resources, common industrial symbiotic platform, and »internal« origin of resources which gives a very good evaluation; proximity is »moderate«; distances are »medium«; the installation/viability of heavy infrastructure is evaluated as »incomplete«; economy and economics are »high«; the reduction of raw material and cost savings, participative management, and cognitive frames are evaluated as »partial«; regulations for waste managements are »fully enforced« (Boshkoska, Rončević and Džajić 2018). The polar map of selected attributes for Barceloneta is given in Fig. 6.4.

According Velenturf and Jensen (2016, 700–709), to move industrial symbiotic research forward there is a need for additional research on the concept of proximity, whilst paying attention to institutional, geographical, cognitive, social, and organizational distances between (Velenturf and Jensen 2016, 700–709).

Kwinana is a »mostly developed« industrial symbiosis because of the accessibility of material resources that are insufficiently available, coming from mixed origin, and the incompleteness of installation/viability of heavy infrastructure.

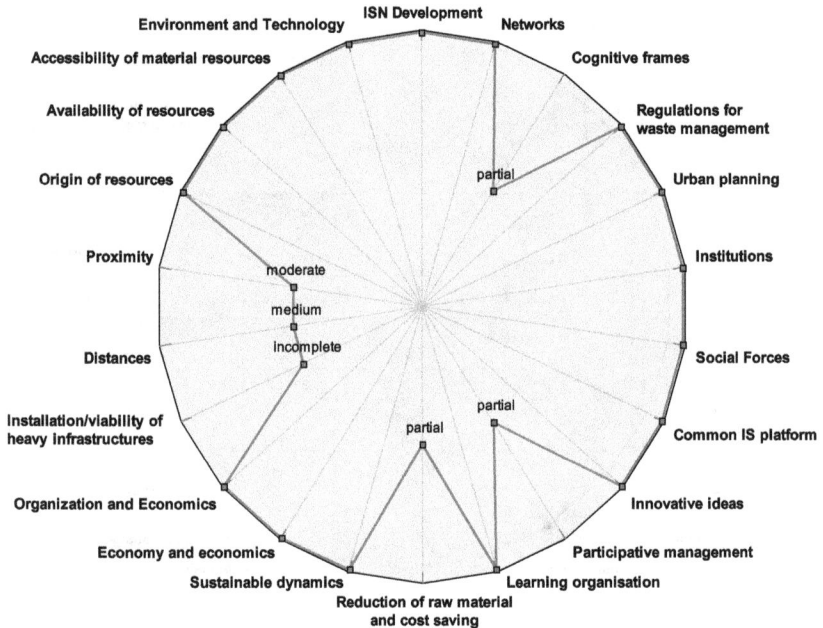

Fig. 6.4: Polar chart of the selected attributes for industrial symbiotic network in Barceloneta. Source: Authors' own calculations in DEXi 2018

The limited competition between operating firms and the isolation of other major industrial centres in eastern Australia are believed to be a cause of such social forces and proximity between actors in Kwinana. There are rigorous regulations in Kwinana, as in Kalundborg, regarding handling the availability of resources to produce alternative fuels (Baas and Huisingh 2008, 399–421; Boshkoska, Rončević and Džajić 2018). The polar map of selected attributes for Kwinana is given in Fig. 6.5.

The evaluation of the attributes are as follows: environment and technology is »weak«; accessibility of material resources is »low«; the availability of resources, sustainable dynamics, the reduction of raw material and cost savings, participative management, innovative ideas, cognitive frames, and networks are »partially/partial«; the origin of resources is »mixed«; proximity and social forces are »moderate«; distances, economy, and economics are »medium«; the installation/viability of heavy infrastructures are »incomplete«; otherwise organization and

Kwinana

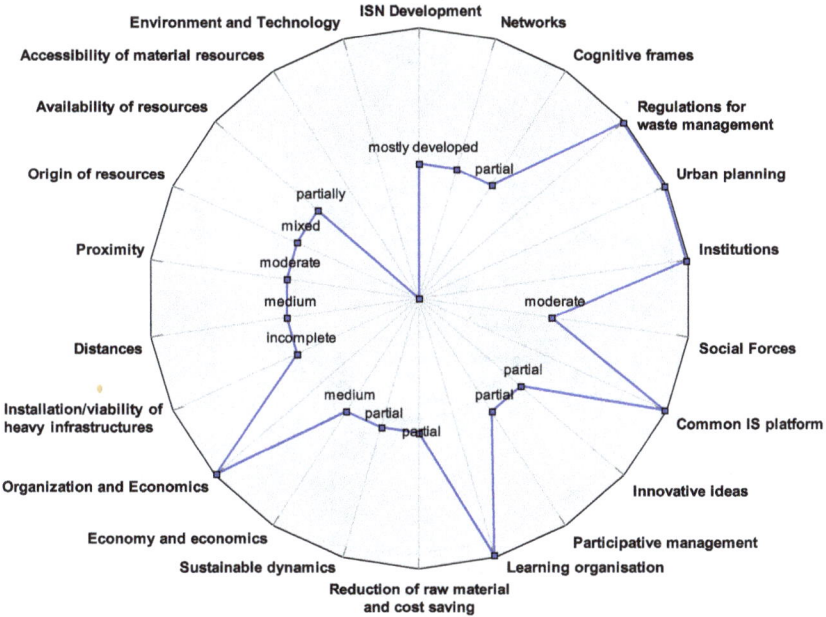

Fig. 6.5: Polar chart of the selected attributes for industrial symbiotic network in Kwinana. Source: Authors' own calculations in DEXi 2018

economics in total are »satisfying« as also are learning organization and urban planning; also there are »strong« forces of institutions and common industrial symbiotic institutes due to »fully enforced« regulations for waste management.

Industrial symbiotic network **Gladstone** is considered »fully developed«. Its strongest aggregate attributes are: environment and technology, learning organizations, social forces except networks, the accessibility of material resources, the origin of resources, and distances (Boshkoska, Rončević and Džajić 2018).

The polar map of the selected attributes for Gladstone is given in Fig. 6.6. The attributes: availability of resources, proximity, installation/viability of heavy infrastructures, organization and economics as economy and economics, sustainable dynamics, reduction of raw material and cost savings, participative management, innovative ideas, and common industrial symbiotic platform were evaluated as partial, medium, or moderate. Here we noted that in the comparison of two of Australia's major heavy industrial regions, Kwinana (Western Australia) and Gladstone (Queensland), we see the different values in

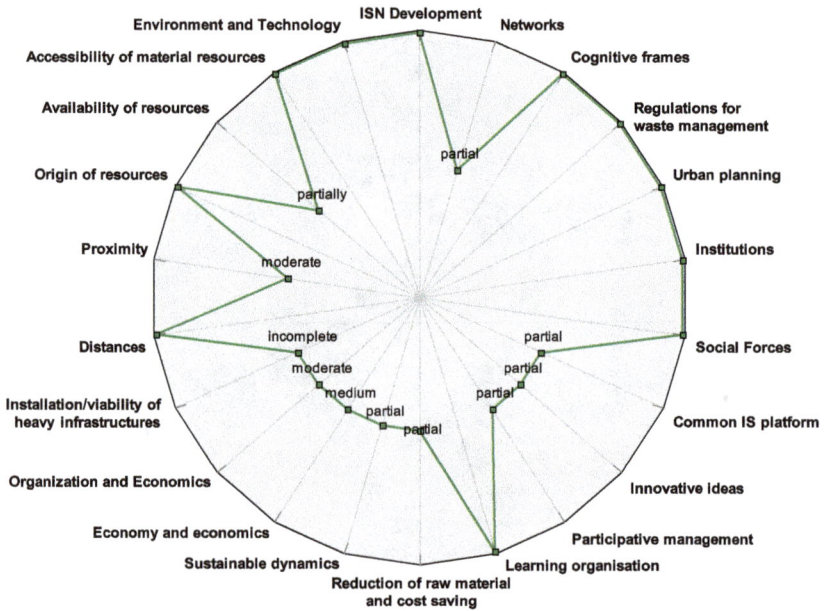

Fig. 6.6: Polar chart of the selected attributes for industrial symbiotic network in Gladstone. Source: Authors' own calculations in DEXi 2018

environment and technology regarding the accessibility of material resources, but the similarity of institutions, urban planning, and regulations (Boshkoska, Rončević and Džajić 2018).

For **Rotterdam** the Industrial symbiotic network development is considered as »mostly developed«. The attributes environment and technology is evaluated as »weak« due to »low« accessibility of material resources. The attribute learning organization is evaluated as »satisfying«, due to the »moderate« values of economy and economics and »partial« common industrial symbiotic platform. The organization and economics are considered »moderate«. We found that some of the »partially enforced« impacts of environmental regulations in Rotterdam are caused by other dimensions of social embedding such as cultural embedding. This means that the effects of informal industrial symbiotic networks on innovative behaviour and on innovation strategies of large firms have been found to be less common in comparison to the reported cases for other industrial symbiotic networks. Despite the dimensions of cognitive frames of

Rotterdam

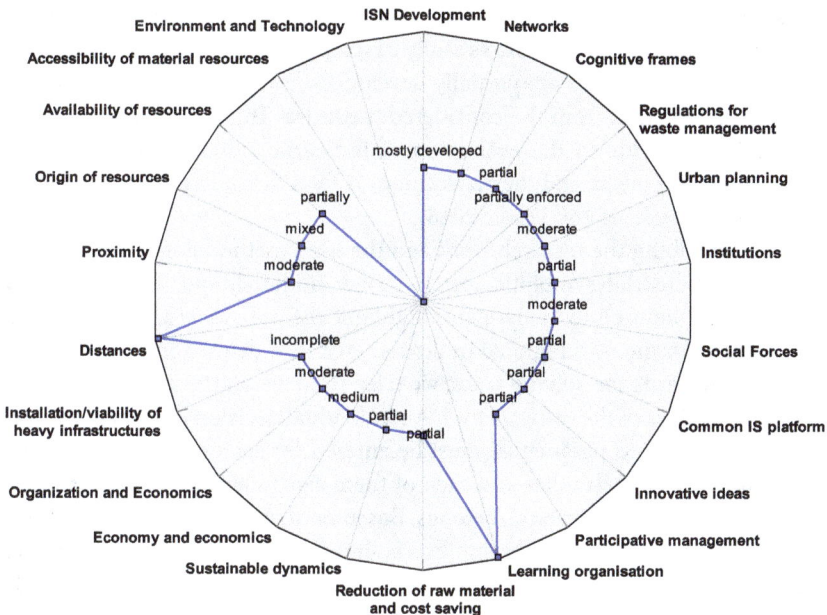

Fig. 6.7: Polar chart of the selected attributes for industrial symbiotic network in Rotterdam. Source: Authors' own calculations in DEXi 2018

embedded routines that are restricting new links in a general (holistic) approach of the Rotterdam Harbour region, there is the possibility for the development of new concepts such as the design of compressed air systems for a group of firms (Baas 2011, 428–440). Social forces are »moderate« due to »partial« institutions, cognitive frames, and networks. Although the Rotterdam Harbour and industry complex is a heavily industrialized area, approximately 100 km² large, with about 1 million inhabitants in the urban surroundings, research projects such as the utilization of reduction waste heat, CO_2 and nutrients from the district heating firms in greenhouses, biofuel synergies, landfill, and urban mining and planning only started in 2010 (Baas 2011, 428–440). Regardless, we (Boshkoska, Rončević and Džajić 2018) evaluated the availability of resources as »partially«, distances as »short«, and the origin of resources and »incomplete« installation/viabilities of heavy infrastructures as »mixed«. The polar map of selected attributes for Rotterdam is given in Fig. 6.7.

In Tab. 6.1 a DEX model is shown and as well as the evaluation of seven indus-trial symbiotic cases. Three of the seven industrial symbiotic cases are evaluated with highest qualitative measure as »fully developed«, while four industrial sym-biotic cases are evaluated as »partially developed«. Such an evaluation is in line with the information from the considered literature. The green coloured words mean a higher value in the evaluation of industrial symbiotic cases, the red colour stands for lower, and the black colour is for a medium evaluated attribute (Boshkoska, Rončević and Džajić 2018).

After describing the research work and the used methodology, the construc-tion of an industrial symbiotic matrix-based model should follow. The new model should be used to analyse and highlight the potentials and barriers that can be found in the new regional industrial area (not defined in this study). As a first step towards the solving tasks, we tried to focus on the user's preferences and the objectives of the software tool. An individual decision maker choses each parameter. Data and parameters must be entered for each case. Parameters are then compared to each other and each of them separately displayed in an area, which, according to set criteria, belongs. Based on these results, the most appro-priate illustration of industrial symbiosis is shown. The problem can occur when we, in the acceptable range, qualify for all cases. Then, it is desirable to review the data, regardless of our wishes, and accurately define the desired limits. Because of the conditional properties of locations, this has not happened. The literature that deal with our idea is limited.

From the results, we found that the Kalundborg case has the best-developed industrial symbiotic networks. This does not exclude the other six cases, because their potential for developing networking in the industrial symbiosis is high.

6.1 Ranking results of utility function

All utility functions that are used in the DEX model are given in Fig. 6.8–6.13. From Fig. 6.8, it follows that the attribute proximity is evaluated as »low« if instal-lation of heavy infrastructures is »impossible« or the distances among firms are »long« and the installation/viability of heavy infrastructures is at most »incom-plete«. Otherwise the proximity is evaluated as »satisfying«.

The evaluation of the attribute Organization and Economics is given in Fig. 6.9. The attribute organization and economics is evaluated as »low« in cases when the economy and economics is »low«, and the learning organiza-tion attribute is evaluated mostly as »moderate«, regardless of the value of the attribute common industrial symbiotic platform. If the attribute common indus-trial symbiotic platform is evaluated as »no« (non-existent) and the learning

Tab. 6.1: DEX model and evaluation of seven industrial symbiotic cases. Source: Authors' own calculation in DEXi 2018

Attribute	Kalundborg	Aalborg	Guayama	Barcelona	Kwinana	Gladstone	Rotterdam
ISN Development	fully developed	mostly developed	mostly developed	fully developed	mostly developed	fully developed	mostly developed
├─ **Environment and Technology**	strong	strong	weak	strong	weak	strong	weak
├─ **Accessibility of material resources**	satisfying	satisfying	low	satisfying	low	satisfying	low
│ └─ Availability of resources	partially	partially	partially	yes	partially	partially	no
│ └─ Origin of resources	internal	internal	mixed	internal	mixed	internal	mixed
├─ **Proximity**	moderate	moderate	moderate	moderate	moderate	moderate	moderate
│ └─ Distances	short	short	medium	medium	medium	short	short
└─ Installation/viability of heavy infrastructures	incomplete	incomplete	incomplete	incomplete	incomplete	incomplete	incomplete
├─ **Organization and Economics**	satisfying	moderate	satisfying	satisfying	satisfying	moderate	moderate
├─ **Economy and economics**	medium	medium	high	high	medium	medium	medium
│ └─ Sustainable dynamics	partial	partial	strong	strong	partial	partial	partial
│ └─ Reduction of raw material and cost saving	partial	partial	partial	partial	partial	partial	partial
├─ **Learning organisation**	satisfying	satisfying	satisfying	satisfying	satisfying	satisfying	satisfying
│ └─ Participative management	partial	strong	partial	partial	partial	partial	partial
│ └─ Innovative ideas	partial	partial	partial	strong	partial	partial	partial
│ └─ Common IS platform	yes	partial	partial	yes	yes	partial	partial
└─ **Social Forces**	satisfying	moderate	moderate	satisfying	moderate	satisfying	moderate
├─ Institutions	strong	partial	partial	strong	strong	strong	partial
│ └─ Urban planning	satisfying	moderate	satisfying	satisfying	satisfying	satisfying	moderate
│ └─ Regulations for waste management	fully enforced	partially enforced	partially enforced	fully enforced	fully enforced	fully enforced	partially enforced
└─ Cognitive frames	strong	strong	partial	partial	partial	strong	partial
└─ Networks	strong	partial	strong	strong	partial	partial	partial

Distances	installation/viability of heavy infrastructures	proximity
29%	71%	
1 long	<=incomplete	low
2 *	impossible	low
3 long	possible	moderate
4 >=medium	incomplete	moderate
5 >=medium	possible	satisfying

Fig. 6.8: Utility function of the aggregated attribute Proximity. Source: Authors' own calculations in DEXi 2018

organization attribute is evaluated as »low«, then regardless of the values of the economy and economics, the attribute organization and economics is evaluated as »low«. It is evaluated as satisfying when economy and economics is evaluated as at least »medium«, while learning organization and common industrial symbiotic platform have their most preferred values of »satisfying« and »yes«, respectively. If economy and economics is evaluated as »high« and learning organization is »satisfying«, then regardless of the value of common industrial symbiotic platform, the organization and economics is evaluated as »satisfying«. However, if common industrial symbiotic platform is »yes« then learning organization should be at least »moderate« so that the organization and economics can be evaluated as »satisfying«. In all other cases, the attribute is moderate (Boshkoska, Rončević and Džajić 2018).

The evaluation of the attribute economy and economics is given in Fig. 6.10. The attribute is evaluated as »low« if at least one of the attributes sustainable dynamics or reduction of raw material and cost savings is evaluated as »weak«, while the other one is evaluated as mostly »partial«. On the other hand, if at least one of the input attributes is evaluated as »strong«, and the other one at least »partial«, then the attribute economy and economics is evaluated with its most preferred value of »high«. In all other cases, the attribute economy and economics is evaluated as »medium« (Boshkoska, Rončević and Džajić 2018).

The evaluation of the attribute learning organization is given in Fig. 6.11. If participative management is »weak« and innovative ideas is at least »partial«, then the learning organization is evaluated as »low«. In the case when the innovative ideas are »weak« and participative management is »partial« then the Learning organization is »moderate«. In all other cases, the attribute is considered to be »satisfying«.

Economy and economics	Learning organisation	Common IS platform	Organization and economics
41%	41%	18%	
1 low	<=moderate		low
2 *	low	no	low
3 low	satisfying	*	moderate
4 <=medium	satisfying	<=partial	moderate
5 medium	<=moderate	>=partial	moderate
6 medium	*	partial	moderate
7 >=medium	low	>=partial	moderate
8 >=medium	<=moderate	partial	moderate
9 medium	moderate	*	moderate
10 medium	>=moderate	<=partial	moderate
11 >=medium	moderate	<=partial	moderate
12 >=medium	satisfying	yes	satisfying
13 high	>=moderate	yes	satisfying
14 high	satisfying	*	satisfying

Fig. 6.9: Utility function of the aggregated attribute Organization and economics.
Source: Authors' own calculations in DEXi 2018

Sustainable dynamics	Reduction of raw material and cost saving	Economy and economics
50%	50%	
1 weak	<=partial	low
2 <=partial	weak	low
3 weak	strong	medium
4 partial	partial	medium
5 strong	weak	medium
6 >=partial	strong	high
7 strong	>=partial	high

Fig. 6.10: Utility function of the aggregated attribute Economy and economics.
Source: Authors' own calculations in DEXi 2018

The evaluation of the attribute social forces is given in Fig. 6.12.. If at least two input attributes are evaluated as »weak«, regardless of the value of the third input attribute, the social forces are evaluated as »low«. Similarly, if at least two input attributes are evaluated as »strong« and the third one is at least »partial«, then social forces are evaluated as »satisfying«. In all other cases the social forces are evaluated as »moderate«.

The evaluation of the attribute institutions is given in Fig. 6.13. Institutions are considered as »strong« if urban planning is considered at least »moderate« and regulations for waste management are »fully enforced«. If urban planning is »low« and regulations for waste management are at most »partially enforced« or urban planning is at most »moderate« and regulations for waste management are »weakly enforced« then institutions are considered as »weak«. In all other cases, institutions are considered as »partial« (Boshkoska, Rončević and Džajić 2018).

	Participative management	Innovative ideas	Learning organisation
	57%	43%	
1	weak	<=partial	low
2	partial	weak	moderate
3	*	strong	satisfying
4	>=partial	>=partial	satisfying
5	strong	*	satisfying

Fig. 6.11: Utility function of the aggregated attribute Learning organization. Source: Authors' own calculations in DEXi 2018

	Institutions	Cognitive frames	Networks	Social forces
	33%	33%	33%	
1	weak	weak	*	low
2	weak	*	weak	low
3	*	weak	weak	low
4	weak	>=partial	>=partial	moderate
5	<=partial	partial	>=partial	moderate
6	<=partial	>=partial	partial	moderate
7	*	partial	partial	moderate
8	partial	<=partial	>=partial	moderate
9	partial	*	partial	moderate
10	>=partial	weak	>=partial	moderate
11	>=partial	<=partial	partial	moderate
12	partial	partial	*	moderate
13	partial	>=partial	<=partial	moderate
14	>=partial	partial	<=partial	moderate
15	>=partial	>=partial	weak	moderate
16	>=partial	strong	strong	satisfying
17	strong	>=partial	strong	satisfying
18	strong	strong	>=partial	satisfying

Fig. 6.12: Utility function of the aggregated attribute Social forces. Source: Authors' own calculations in DEXi 2018

	Urban planning	Regulations for waste management	Institutions
	43%	57%	
1	low	<=partially enforced	weak
2	<=moderate	weakly enforced	weak
3	low	fully enforced	partial
4	>=moderate	partially enforced	partial
5	satisfying	<=partially enforced	partial
6	>=moderate	fully enforced	strong

Fig. 6.13: Utility function of the aggregated attribute Institutions. Source: Authors' own calculations in DEXi 2018

6.2 Supposition and expectation of the results

The analysis of results leads us to the following assumptions: the construction of our new model has been possible by following meta-analysis to »guess« the cause-effect relationships, in regards to the critically realistic philosophical perspective. When trying to construct a model and »reduce« the perceived reality of industrial symbiotic systems, the research scope places great importance on social interactions. The social interactions are assumed to lay the foundations to create synergies between firms. This would be undertaken at two different levels: (1) in the methods of data gathering by using other sources of information, mainly secondary material about the network and actors and, (2) in the preparation and selection, by contrasting and cross-comparing the secondary information that were used as input data into the software tool DEXi.

Hence, the hypothesis, which states that: »Industrial Symbiosis is not only a paradigm with important technological, economic, and environmental consequences. Industrial symbiosis is more than environmentally sustainable industrial activity. Industrial symbiosis is also a social paradigm and its manifestations in the form of social networks are an indistinguishable part of the development process of »Industrial Symbiotic Networks«. Is largely confirmed and the results from the evaluation of industrial symbiotic networks suggest to the decision makers which particular spheres and which attributes need to be improved to develop or obtain a better industrial symbiotic network« (Boshkoska, Rončević and Džajić 2018).

In practice, the model could also help regional firms, local and state institutions, and the public sector, in understanding the weaknesses of the current system, and to define weak spots for which solutions are required. The model can also support the decision makers in the process of improving the industrial symbiotic networks and its temporal as well as long-term behaviour (Criado Pacheco et al. 2017). However, the following research questions remain open for further research: What kinds of indicators are appropriate and suitable to measure the development of industrial symbiotic networks? How can the development of industrial symbiotic networks be initiated in order to add value to unused resources in our new industrial area?

This developed model can be used as an analysis tool to evaluate the degree of compliance of our new industrial area with the industrial symbiotic prerequisites. It will then be allowed to determine what are the facilitating mechanisms already existing, what are the barriers to avoid, but also emit recommendations regarding future operations to be undertaken. A more implicit objective would be, through the development of such study, to create »awareness«, known to be important triggering mechanisms, among the structures and actors of our model of an industrial park (Boshkoska, Rončević and Džajić 2018).

In the presented industrial symbiotic cases, it has been seen that managers from the same industrial area may interact with each other by extra-professional social activities (associations, clubs, sport activities), that lead to deepened collaborations by enhancing trust between managers, in a professional context. However, the mechanisms of willingness for mutual cooperation are not clear: a major part sees their input/output flows as confidential, and another major part indicates mutual confidentiality and loyalty as imperative rules to be respected. It may be possible to see a link between the lack of willingness and certain nervousness in regards to sharing business-sensitive information. An isolated industrial area can ensure a collaborative atmosphere between the operating firms.

As revealed from industrial symbiotic case studies«, firms located near to each other tend to observe »the next-door neighbours«, maintain social links with them, and eventually share points of interest and development projections. It seems even though there are almost not any existing collaborations now, firms roughly tend to know each other, including core activity and input/output material flows. It is believed that the poor distance separating firms and the small size of the overall area induce such degrees of intimacy. Here are some conditions/prerequisites of the firms' collaborations found in the literature about industrial symbiosis:

- The firm should be given access to technology and be aware of it.
- The firm must be a certain economic, social, and political power to be able to invest in the technology.
- The firm interested in introducing the technology must be present in the society.
- The firm must have a clear organizational structure in order to be able to make a decision.
- The firm must acquire the knowledge to be able to handle and maintain the decision-making processes (Kalundborg Symbiose 2017).

Nowadays, firms are increasingly aware of the need to come up with innovative solutions in terms of products, to reduce their impact on the environment, or to support their business strategy. A participative management is an appropriate way to deal with this problem and bring up innovation. This fostering mechanism is considered as another brick in the industrial symbiosis wall development process. The literature relevant to presented industrial symbiotic cases resulted in a predominantly bottom-up participative management style within firms. It is thought that it can have a beneficial influence by supporting employees' ideas in regards to possible improvements or innovative solutions. From that reason, it is believed that firms should take into consideration employees' opinions to successfully start and develop industrial symbiotic partnerships.

DiMaggio and Powell (1991) distinguish several mechanisms of transmission in the organizational field, which have an effect on a social field. A field is defined by the interaction among its actors because they mutually acknowledge being involved in similar/related activities. The boundaries of the field are thus an empirical variable. Building on this, the following mechanisms of transmission can be proposed: private interest government, imitation, training and professionalization, demonstration projects, and altering boundary conditions (Boons and Baas 2006).

These mechanisms may all play a role in the diffusion of industrial symbiotic networks concepts and routines throughout a society. A major initial condition is the way in which actors are linked at this level, that is, through what ways firms and coordinating actors in regional industrial clusters can acquire information in the larger society about industrial symbiotic networks and have access to the resources to start implementing them. Here we may find considerable differences. These differences, for example, in one industrial symbiotic case combine into approaches that are distinct for societies, as research on national developments in the UK (National Industrial Symbiosis Programme), the Netherlands (sustainable industrial parks), the US (and the Presidential Council on sustainable Development made in 1996), China (circular economy), and Japan (urban symbiosis) indicate. The investigation revealed no symbiotic linkages and, other than a few agglomerations benefits aside from a common plot of property and basic utilities, a very low threshold of static urbanization economies achieved passively by co-located firms. It was found that most managers of firms in the eco-industrial parks did not even know each other, and without these social ties the results seen in some industrial symbiotic case were not possible. Even serious common problems, such as seasonal flooding that made roads in the eco-industrial parks impassable, were not being jointly addressed (Johnson et al., 2003 quoted in Chertow, Weslynne and Espinosa 2008, 1308).

At the societal level, actors may be involved in the diffusion of industrial symbiotic networking, including the underlying philosophy and rationale (Czarniawska and Sevón 1996; DiMaggio and Powell 1983). The theory of the diffusion deals with the mechanisms that are responsible for the diffusion of a concept, innovation, or idea among a set of organizations. This is indicated by the adoption of the concept in language (corporate newsletters and internal memoranda, annual reports, descriptions towards actors) and in practice, through the adoption of certain organizational routines. Diffusion can be a result of transmission of a concept from one organization/individual to another, or by a process of selection through which organizations that do not use the concept are eliminated (Boons and Howard-Grenville 2009).

In the industrial symbiotic cases proposed in this study, it is documented that the business case for synergies does not depend solely on the financial benefits, but also it is affected by other sustainability aspects, such as risk management, continued access to vital resources, environmental legislation, and community relations.

The assessment of drivers, barriers, and triggers for regional synergy developments appear to fall into six broad categories: economics, information availability, organizational and social issues, region-specific issues, regulations, and technical issues. From this assessment, a wide range of drivers and barriers exist, which are influenced by a diverse set of actors (e.g. firms, regulators, community) (van Beers et al. 2007a). The complete set of drivers, barriers, and trigger events, rather than one specific aspect, determine the business and sustainability case for a regional synergy opportunity and hence are key to the implementation of by-product or utility synergy projects (van Beers et al. 2007b, 55–72). Overall, there is no »one-size-fits-all« approach to developing regional synergies; »each synergy is unique in terms of its drivers, barriers, business case and sustainability benefits« (van Beers et al. 2007a, 830–841).

Firms communicate with each other, which can happen via the help of a common communication platform, to keep a comprehensive vision of the whole system.

Within the extent of their knowledge, firms are, most of the time, interacting in accordance to economical profitability maximization. In this way, local governments seem interested in the idea of developing an industrial symbiotic organization within the city. It is critical to remember the importance of public neutrality in decision-making processes. Therefore, local governments should come as a support and should not try to take control over firms' decisions. As mentioned in the previous section, firms are most likely the best actors in position to receive market price information and to make decisions in accordance. It is important that firms do not feel forced to undertake synergies from an above authority and conserve the ownership of the initiative.

From the literature, we found that is important to activate and encourage the participation of the local governments in such a project in several forms: it must legislate flexible regulations in order not to slow down synergies possibilities, it must promote the industrial symbiotic development in the surroundings to attract investors, it must allow infrastructures crossing public areas, etc.

The role of power is hardly discussed systematically here. Maybe this should be done, as it is one of the more difficult concepts of sociology in terms of empirical analysis. Nevertheless, actors are not equally able to influence each other's actions and system outcomes and this should be considered. Especially the political, cognitive and cultural forms of embeddedness determine the role of political-economic systems, ideologies and social networks in both enabling and constraining voluntary action in a variety of ways (Beckert 2009, 251).

The structure of this study then specifies basic industrial symbiotic relationships and establishes the relationships amongst them.

Sociological approaches to the economy explain economic outcomes based on the influence of social structures on individual action, economy explains economic outcomes based on the influence of social structures on an individual act. By explaining the resolution of coordination problems and distributional outcomes by means of the social forces entangling market actors, sociology provides an alternative to those economic approaches that proceed on the basis of the individual interests of actors when addressing the question of economic order.

The question is whether industrial ecology/industrial symbiosis and later networks are part of the sustainability system. The number of firms, their diversity in size and type, and the intensity of their interactions are major variables in the system. Here the links between individual firms and the links between firms and society are to be tested according to the criteria of sustainability. This system demands an integrated approach based on new worldviews. The production process is an element at the level of individual firms (at the micro level) but the output of by-products is also the function of the servant of the network (Wallner 1999, 49–58) at the mezzo level.

»Several social science characteristics of industrial ecology efforts are identified, among which are: embeddedness, capabilities, status quo, reform, or transformation beyond the status quo (Baas and Huising 2008, 400–412). The different dimensions can be described in terms of cognitive, structural, cultural, political, spatial and temporal embeddedness« (Baas and Huising 2008, 400–412). Simsek, Lubatkin, and Floyd (2003) use also relational embeddedness as a reference when they describe the quality of the link, underlining the effects of reliable ties between social actors on their economic activities. They link relational embeddedness strongly to organizational (Granovetter 1992, 3–11; Gnyawali and Madhavan 2001, 431–445) and rational embeddedness. Actors itself, far from being »atomistic entities« free to undertake any competitive action within their own resource limitations, are embedded in a network of relationships that influences their competitive behaviour. The structure of the network to which they belong to influences the flow of assets, information, and status among the network actors. Network-based resource advantages vary across firms, resulting in varied levels of motivation and ability to undertake action, or respond to the actions of others. Firms can work together by sharing resources and committing themselves to common task goals in certain domains; at the same time, partners also compete by taking independent positions in other domains.

This social work leads to a necessary cohesion in order to develop further willingness to invest into industrial symbiotic synergies. When such a situation is reached, the following interactions between actors are so that there are social interactions between firms, especially between managers in order to maintain good communication, trust, and willingness when processing operations involving sensitive business information and environmental risks regarding the handling of waste. There may be interactions between operating firms and a third party, such as the municipality or scholars. It happens when the third party tries to bring education, awareness, and training to operating firms, in order to transform the latter entities into actual social carriers of techniques for industrial symbiotic development, as stressed in the paper by Edquist and Edqvist (1979, 313–331).

The two-way arrows between the three spheres indicates a cyclic development. A strong social cohesion is thought to lead to an organizational development between actors, which sets the necessary elements to achieve synergies in terms of technical feasibility. In turn, a strong and viable synergistic network may reinforce social proximity between actors and often brings new actors in the network, which increases experience and knowledge at the organizational level and eventually increases physical possibilities of synergies.

It is thought that, at an early stage of development, the order is likely to be Social-Organizational-Physical. However, the outcome resulting in the physical establishment of a cooperation might benefit the organizational level as well as the social level. This »learning-by-doing« process does not necessarily follow the Social-Organizational-Physical cyclic pattern. However, as aforementioned, it is assumed that in the first round, primary social work needs to be done to be able to organize a potential partnership before physically realizing it.

This »learning-by-doing« assumption is confirmed by a paper work by Harris and Pritchard (2004, 89–111) dealing with learning organizations in an industrial symbiotic context: Successful exchanges would then be presented at network meetings potentially boosting more and building on the learning of the network.

7 Conclusion and discussion

Let us briefly summarize our key findings first. In the second chapter, we attempted to understand how the existing literature explains industrial ecology, industrial symbiosis. We conceptualized the morphogenesis/development of industrial symbiotic networks. We theorized industrial symbiotic's social aspects into structure of contemporary theories of sociology of social fields, development, behaviour, values, and principles in the industrial symbiotic networking. Seven industrial areas were exanimated and their main features were outlined. The methodological part presented a qualitative multi-criteria model for evaluating the level of developed regional industrial symbiotic networks. The model is based on the available literature for industrial symbiosis and its networks, from which we have extracted the most relevant and most general attributes to build it. We have also defined the relations amongst attributes in terms of easily understandable »if... then« rules. We outlined the key benefits of the model of industrial symbiotic network, we supported our statements using seven of the most famous examples of industrial symbiotic best practice, and we provided a detailed description of the attributes and utility functions of the aggregated attributes (Boshkoska, Džajić and Rončević 2018).

In Chapter 6, we presented how the values of the attributes were obtained and how difficult it is to measure each attribute quantitatively. A design through a set of indicators from each of the spheres was proposed, which has a comparative and/or descriptive role to understand the requirements of the development of an industrial symbiotic network.

In Chapter 7, since the evaluation of industrial symbiotic networks based on one qualitative indicator leads to a high uncertainty in the evaluations, we used an approach in which we developed several indicators for the input attributes in the model. Based on the number of indicators that the industrial symbiotic network fulfils for the given attribute, we performed a more realistic evaluation of industrial symbiotic networks. Still, the evaluations could be changed by adopting new literature sources. Our belief with the use of the new model is that the decision makers could surpass the imposed disciplinary boundaries of the involved interdisciplinary research areas. This would enable them to evaluate and analyse the requirements for developing improved industrial symbiotic networks. Limited management attention given to the industrial symbiotic network concept was a common result found through the study of industrial symbiotic cases. These can be overcome by helping decision makers understand the importance of the development of industrial symbiotic networks and the proposed model could be used to provide support to the decision makers in this context. So, the next challenge is identifying opportunities for creating regional industrial symbiotic networks. The creation of local dynamics and platforms that intensify communication among actors is one effective way of addressing this challenge. Further support can be provided through the systemic assessment of the needs and capacities of all actors included in the process of the development of a local industrial symbiotic network.

Recently, however, theoretical sociology contributions to networks and networking in industrial symbiosis have uncovered a lot of barriers for increased industrial symbiotic activities. One of the baggiest barriers nowadays is still the lack of awareness about the potentially strategic business and development of the value of industrial symbiosis.

On the basis of literature analyses and concrete industrial symbiotic examples, we have tried to define and evaluate as closely as possible the attributes of these cases and attributes in order to establish a new model. We have also tried to find out how existing literature explains the phenomena of the complex process of social aspects and the rise of networking in the industrial symbiosis. In addition, we have tried to concentrate also on other explanatory reasons; we used sociological category of »organization« or »structure«, which refers to everything affecting the action of the actors, both of a physical nature (e.g. networks and infrastructure) and of a purely social nature (standards, rules, behavioural models, organizational procedures, routines, power configurations, dominant cognitive models and representations, etc.); and »social fields«, which refers to the level of social integration, expressed in aspects such as the degree of *mutual trust* among citizens and between citizens and public authorities, and the cognitive capital, networks, and institution. How social fields change are assessed is explained by Jens Beckert (2009): The different approaches of economic sociology develop separately from each other is, nevertheless, only partially justified. While some authors have attempted to eradicate the social structures discussed by competing approaches by claiming that they do not have independent effects, many others have attempted to integrate them by considering them simultaneously.

Cognitive frames have been brought into institutional theory by combining them with networks and institutions. This can be seen in sociological institutionalism, which emphasizes the role of cognitive frames, meaning structures as decisive for the explanation of economic outcomes by broadening the notion of institution; institutions are defined as intersubjectively shared meanings and thereby become almost indistinguishable from cognitive frames. A similar strategy can also be detected in those approaches of networking that attempt to »endogenize« understanding-to develop something internally in an eco-nomic view. Networks are interpreted as networks of meaning« and are said to exercise their influence on actors based on narratives which express mental maps of the structure of social relations. Hence, in these accounts, the objectivity of networks is not constituted by the position of relations and their connections as such, but by the dominant interpretations through which actors perceive the industrial symbiotic networks. Some experts in net-working conceive of institutions as »congealed networks where interactions between people gradually acquire an objective quality and eventually people take them for granted. In this manner, institutions are integrated into the approach, but in a form that makes them indis-tinguishable from networks. What is at liability in these strategies is that they are conflating the different types of social forces and therefore fail to take into consideration their analytic disconnection« (Beckert 2010, 607).

This study follows the hypothesis and in the development of industrial symbiotic networks, it considers all three structural forces together.

»Another interesting mechanism was found: in relatively sized industrial symbiotic networks of a few dozen partners/firms, the means for trust building include personal meetings, regular communication, the sharing of information and knowledge, and stable rules of the game, whereas computational means can also be beneficial to support trust« (quoted in Alkhalil 2016, 318). Establishing trust among industrial symbiotic actors is an important mechanism in collab-orative projects, since the possibility of opportunistic behaviour of partners cannot be eliminated by formal contracts.

As van Beers et al. (2007a) discussed, social networking and interaction is believed to be a pivotal mechanism when building up interests, trust, and commitments between the actors of an industrial symbiotic system and firms are not forced to cooperate with others. Although economical profitability is a strong incentive pushing cooperation, it does not cover risks relating to confidentiality issues. However, we have shown that the social dimen-sion of analysis encompasses social and human phenomena that enhances the development of mechanisms (trust, willingness, and ties), generating effects in favour of industrial symbiotic networks. This emphasizes mainly the human side of such system; as social interactions between human beings (e.g. managers), but people as social forces, networks, and cognitive frames, are seen as having a bilateral agreement framing synergies that imply the need to know each other's production processes, waste, quantity and quality of mate-rial required, and technology used. These data represent sensitive information that might put firms at risk if publicly known by competitors. Therefore, it is believed that social interactions gather effects such as trust, good communi-cation, and proximity, required to cover confidentiality issues related to risks.

This is especially true when considering the specific nature of waste being exchanged: Communication and trust are thought to be important when the materials being exchanged have potential disadvantages, because they are subject to environmental regulations (Chertow 2004). As we may note in the whole study, communication and trust among actors/stakeholders/ managers in the industrial symbiosis are thought to play important roles in exchanges of waste; for this reason, such analyses and methodology like this, based on social networks and the implementation of software tool, can easily show and measure industrial networking in industrial symbiosis. In this way, we can identify the prevalence of industrial symbiotic linkages. The study quantified patterns in various relationships among firms and net-working, including technological, logistic, economical, and social aspects, through supply chains and informal ones through interpersonal interactions. Outlines in relationships or social structure result from constant collabora-tion among actors and can be represented as networks (Ashton 2008, 34–51). A network consists of representing the actors in the system and links indi-cating various connections between them. Actors may be discrete social units, such as individuals, or collective entities, such as organizations (Wasserman and Faust 1994). Also, the willingness, sometimes called the »human factor«, refers to the human feeling of desire to invest into an industrial symbi-otic synergy project. Willingness to invest in synergies with other firms is seen to be a prerequisite when implementing industrial symbiosis, since it is formed by having an awareness of industrial symbiotic benefits, shared technical knowledge, trust, and good communication with one another. Although the Mechanisms, Cognitive frames, and Networks have origi-nally the same causes, it is believed that the former leads to human-oriented effects, while the Social forces are more industrial symbiotic–oriented effects.

We largely confirmed the hypothesis, based on a general methodological ap-proach which is expressed in three basic methodological choices:

- The choice of using, as an empirical basis, existing knowledge produced by researchers or held by the energy transition of players, already available or obtainable through secondary data (documentary sources of different degrees of formalization), position papers, articles, scientific literature, projects, websites, blogs, policy planning documents, research reports, statistical reports, policy makers, the scientific community, experts, and professionals.
- The decision to recognize the same heuristic importance to all the phenomena identified, regardless of whether it has been identified through quantitative or qualitative information or through operational or cognitive phenomena.

- The decision to include in this empirical basis not only information on operational phenomena (events, facts, or actions of any kind), but also data on cognitive phenomena, expressions of orientations and the intent of energy transition actors (Džajić and Rončević 2017).

Industrial symbiosis has been found to be driven not only by economic considerations, such as lowering costs for waste disposal, as well as environmental considerations, such as accessing limited natural resource supplies, but social problems such as institutions (laws, central concern for law, the formal mechanism for political rule making and enforcement); cognitive frames (social interaction, meaning-making technologies, and strategically-selective opportunities for reflection and learning); and social networking (social structure made up of a set of social actors such as individuals or organizations) – a set of the dyadic ties between these actors plays an important role (Beckert 2009).

In the world of physics, obviously, things do not remain at rest or in uniform motion all the time. In fact, all kinds of interesting things happen. The way this is accommodated in physical law is to say that something called »forces« act on particles or masses and cause them to change the way they are moving. In a very simple analogy, it is proposed that there are various kinds of cognitive and informational forces, which act on human activities and cause to change their behaviour or motion. In the physical analogy, external forces cause these changes. In the case of humans searching, it is proposed that they be thought of as transitions, caused by or occurring in response to external events.

The history and the development of the industrial symbiosis helps us to explain how industrial symbiotic networks can be rationalized. By adopting a systemic approach through methods of industrial symbiotic cases, we reveal several fostering mechanisms in a model on how to start and embed industrial symbiosis in the environment where such a phenomena is unknown. The integration of economic, environmental, and social dimensions in industrial activities is increasingly perceived as a necessary condition for a sustainable society. A variety of methodological approaches and processing techniques are set up: theory (to build clusters of homogeneous phenomena in terms of fundamental characteristics), graph theory (to identify connections between phenomena or phenomena clusters), semantics (to manage the findings of a linguistic nature through semantic condensation processes or the construction of separate semantic fields), and systems theory (especially to isolate the different components of the phenomena and processes analysed and to detect any functional relationships between them).

The uniqueness of this research is the developed qualitative DEX model, which provides a quick overview over the development of industrial symbiotic networks over three different areas. This can be achieved through the development of social capital between industries, communities, and government. The novelty is in the developed DEX model, which is easy to understand not only for researchers in one of the industrial symbiotic network areas discussed in the paper, but it can also be easily understood and interpreted by the public and decision makers. The model is based on the industrial symbiotic studies in the literature and reflects the current state of industrial symbiotic networks as described in the literature. We believe that industrial symbiosis is a much more complex phenomena as we have discovered and presented. It is difficult to explain its development through a set of indicators, but the model developed can potentially have a descriptive (or comparative) role in understanding the dynamics of industrial symbiosis and industrial symbiotic network development. Hence, this is the first DEX model, which may be developed upon further. However, we wanted it to be as simple as possible for the audience at this phase, so that it can be used as a general model that incorporates the most important views of an industrial symbiotic network (organization, economics, and social views) (see all in Boshkoska, Rončević and Džajić 2018).

That social beliefs, values, and norms develop within a social system and influence an organization's/firms' behaviour and function inside and outside. It is expected that new norms will emerge among industrial symbiotic field. There are different types of relationships amongst firms in industrial symbiosis, and within those firms, most relationships are mediated through a few actors. The social structural forces in the industrial symbiosis exist and are presented and studied in this study (from developed model) as: values, culture, ethics, and morals, which play a very important role in constructing industrial symbiosis and its networks.

Appendix

Attribute tree

Attribute	Description
ISN Development	Current ISN Development
⊢**Environment and Technology**	Facets, location, diversity of industries, presence of river, sea, core businesses
| ⊢**Accessibility of material resources**	Scarcity of resources
| | ⊢Availability of resources	Industrial water, energy, production materials (primary and secondary)
| | └Origin of resources	How the available resources are connected: internal, external, mixed
| └**Proximity**	Physical distance among industrial installations
| ⊢Distances	Distances among firms
| └Installation/viability of heavy infrastructures	Vicinity of the main infrastructure suppliers to the IS area
⊢**Organization and Economics**	Economic benefits, profitability
| ⊢**Economy and economics**	Yielding the benefits of industrial symbiosis
| | ⊢Sustainable dynamics	Create economic growth, while advancing social and environmental objectives
| | └Reduction of raw material and cost saving	Reduction, reuse and recycling; Income generating synergies
| ⊢**Learning organisation**	An approach to include innovative ideas of employees in the decision making process
| | ⊢Participative management	Involvement of employees into decision making processes
| | └Innovative ideas	Employees generate innovative ideas
| └Common IS platform	Existence of a joint and comprehensive policy and communication platform
└**Social Forces**	Social factors shaping the topography of IS networks
⊢**Institutions**	Rules, ritualised behaviours
| ⊢Urban planning	Local and/or regional spatial planning
| └Regulations for waste management	Implementation of the waste management regulations
⊢Cognitive frames	Green values, public awareness and green procurement
└Networks	Social trust and ease of cooperation

Scales

Attribute	Scale
ISN Development	partially developed; mostly developed; *fully developed*
⊢**Environment and Technology**	weak; moderate; *strong*
| ⊢**Accessibility of material resources**	low; moderate; *satisfying*
| | ⊢Availability of resources	no; partially; *yes*
| | └Origin of resources	external; mixed; *internal*
| └**Proximity**	low; moderate; *satisfying*
| ⊢Distances	long; medium; *short*
| └Installation/viability of heavy infrastructures	impossible; incomplete; *possible*
⊢**Organization and Economics**	low; moderate; *satisfying*
| ⊢**Economy and economics**	low; medium; *high*
| | ⊢Sustainable dynamics	weak; partial; *strong*
| | └Reduction of raw material and cost saving	weak; partial; *strong*
| ⊢**Learning organisation**	low; moderate; *satisfying*
| | ⊢Participative management	weak; partial; *strong*
| | └Innovative ideas	weak; partial; *strong*
| └Common IS platform	no; partial; *yes*
└**Social Forces**	low; moderate; *satisfying*
⊢**Institutions**	weak; partial; *strong*
| ⊢Urban planning	low; moderate; *satisfying*
| └Regulations for waste management	weakly enforced; partially enforced; *fully enforced*
⊢Cognitive frames	weak; partial; *strong*
└Networks	weak; partial; *strong*

ISN Development

Current ISN Development

1. partially developed
2. mostly developed
3. fully developed

Environment and Technology

Facets, location, diversity of industries, presence of river, sea, core businesses

1. weak
2. moderate
3. strong

Accessibility of material resources

Scarcity of resources

1. low
2. moderate
3. satisfying

Availability of resources

Industrial water, energy, production materials (primary and secondary)

1. no
2. partially
3. yes

Origin of resources

How the available resources are connected: internal, external, mixed

1. external
2. mixed
3. internal

Proximity

Physical distance among industrial installations

1. low
2. moderate
3. satisfying

Distances

Distances among firms

1. long
2. medium
3. short

Installation/viability of heavy infrastructures

Vicinity of the main infrastructure suppliers to the IS area

1. impossible
2. incomplete
3. possible

Organization and Economics

Economic benefits, profitability

1. low
2. moderate
3. satisfying

Economy and economics

Yielding the benefits of industrial symbiosis

1. low
2. medium
3. *high*

Sustainable dynamics

Create economic growth, while advancing social and environmental objectives

1. weak
2. partial
3. *strong*

Reduction of raw material and cost saving

Reduction, reuse and recycling; Income generating synergies

1. weak
2. partial
3. *strong*

Learning organisation

An approach to include innovative ideas of employees in the decision making process

1. low
2. moderate
3. *satisfying*

Participative management

Involvement of employees into decision making processes

1. weak
2. partial
3. *strong*

Innovative ideas

Employees generate innovative ideas

1. weak
2. partial
3. *strong*

Common IS platform

Existence of a joint and comprehensive policy and communication platform

1. no
2. partial
3. *yes*

Social Forces

Social factors shaping the topography of IS networks

1. low
2. moderate
3. *satisfying*

Institutions

Rules, ritualised behaviours

1. weak
2. partial
3. *strong*

Urban planning

Local and/or regional spatial planning

1. low
2. moderate
3. satisfying

Regulations for waste management

Implementation of the waste management regulations

1. weakly enforced
2. partially enforced
3. fully enforced

Cognitive frames

Green values, public awareness and green procurement

1. weak
2. partial
3. strong

Networks

Social trust and ease of cooperation

1. weak
2. partial
3. strong

Functions

Attribute	Rules	Defined	Determined	Values
ISN Development	27/27	100,00%	100,00%	partially developed:7,mostly developed:16,fully developed:4
├─Environment and Technology	9/9	100,00%	100,00%	weak:4,moderate:3,strong:2
│ ├─Accessibility of material resources	9/9	100,00%	100,00%	low:6,moderate:1,satisfying:2
│ │ ├─Availability of resources				
│ │ └─Origin of resources				
│ └─Proximity	9/9	100,00%	100,00%	low:4,moderate:3,satisfying:2
│ ├─Distances				
│ └─Installation/viability of heavy infrastructures				
├─Organization and Economics	27/27	100,00%	100,00%	low:8,moderate:14,satisfying:5
│ ├─Economy and economics	9/9	100,00%	100,00%	low:3,medium:3,high:3
│ │ ├─Sustainable dynamics				
│ │ └─Reduction of raw material and cost saving				
│ ├─Learning organisation	9/9	100,00%	100,00%	low:2,moderate:1,satisfying:6
│ │ ├─Participative management				
│ │ └─Innovative ideas				
│ └─Common IS platform				
└─Social Forces	27/27	100,00%	100,00%	low:7,moderate:16,satisfying:4
├─Institutions	9/9	100,00%	100,00%	weak:3,partial:4,strong:2
│ ├─Urban planning				
│ └─Regulations for waste management				
├─Cognitive frames				
└─Networks				

Tables

	Environment and Technology	Organization and Economics	Social Forces	ISN Development
	33%	33%	33%	
1	weak	low	*	partially developed
2	weak	*	low	partially developed
3	*	low	low	partially developed
4	weak	>=moderate	>=moderate	mostly developed
5	<=moderate	moderate	>=moderate	mostly developed
6	<=moderate	>=moderate	moderate	mostly developed
7	*	moderate	moderate	mostly developed
8	moderate	<=moderate	>=moderate	mostly developed
9	moderate	*	moderate	mostly developed
10	>=moderate	low	>=moderate	mostly developed
11	>=moderate	<=moderate	moderate	mostly developed
12	moderate	moderate	*	mostly developed
13	moderate	>=moderate	<=moderate	mostly developed
14	>=moderate	moderate	<=moderate	mostly developed
15	>=moderate	>=moderate	low	mostly developed
16	>=moderate	satisfying	satisfying	fully developed
17	strong	>=moderate	satisfying	fully developed
18	strong	satisfying	>=moderate	fully developed

	Accessibility of material resources	Proximity	Environment and Technology
	71%	29%	
1	low	*	weak
2	<=moderate	low	weak
3	moderate	>=moderate	moderate
4	satisfying	low	moderate
5	satisfying	>=moderate	strong

	Availability of resources	Origin of resources	Accessibility of material resources
	43%	57%	
1	no	*	low
2	<=partially	<=mixed	low
3	*	external	low
4	yes	mixed	moderate
5	>=partially	internal	satisfying

	Distances	Installation/viability of heavy infrastructures	Proximity
	29%	71%	
1	long	<=incomplete	low
2	*	impossible	low
3	long	possible	moderate
4	>=medium	incomplete	moderate
5	>=medium	possible	satisfying

	Economy and economics	Learning organisation	Common IS platform	Organization and Economics
	41%	41%	18%	
1	low	<=moderate	*	low
2	*	low	no	low
3	low	satisfying	*	moderate
4	<=medium	satisfying	<=partial	moderate
5	medium	<=moderate	>=partial	moderate
6	medium	*	partial	moderate
7	>=medium	low	>=partial	moderate
8	>=medium	<=moderate	partial	moderate
9	medium	moderate	*	moderate
10	medium	>=moderate	<=partial	moderate
11	>=medium	moderate	<=partial	moderate
12	>=medium	satisfying	yes	satisfying
13	high	>=moderate	yes	satisfying
14	high	satisfying	*	satisfying

	Sustainable dynamics	Reduction of raw material and cost saving	Economy and economics
	50%	50%	
1	weak	<=partial	low
2	<=partial	weak	low
3	weak	strong	medium
4	partial	partial	medium
5	strong	weak	medium
6	>=partial	strong	high
7	strong	>=partial	high

	Participative management	Innovative ideas	Learning organisation
	57%	43%	
1	weak	<=partial	low
2	partial	weak	moderate
3	*	strong	satisfying
4	>=partial	>=partial	satisfying
5	strong	*	satisfying

	Institutions	Cognitive frames	Networks	Social Forces
	33%	33%	33%	
1	weak	weak	*	low
2	weak	*	weak	low
3	*	weak	weak	low
4	weak	>=partial	>=partial	moderate
5	<=partial	partial	>=partial	moderate
6	<=partial	>=partial	partial	moderate
7	*	partial	partial	moderate
8	partial	<=partial	>=partial	moderate
9	partial	*	partial	moderate
10	>=partial	weak	>=partial	moderate
11	>=partial	<=partial	partial	moderate
12	partial	partial	*	moderate
13	partial	>=partial	<=partial	moderate
14	>=partial	partial	<=partial	moderate
15	>=partial	>=partial	weak	moderate
16	>=partial	strong	strong	satisfying
17	strong	>=partial	strong	satisfying
18	strong	strong	>=partial	satisfying

	Urban planning	Regulations for waste management	Institutions
	43%	57%	
1	low	<=partially enforced	weak
2	<=moderate	weakly enforced	weak
3	low	fully enforced	partial
4	>=moderate	partially enforced	partial
5	satisfying	<=partially enforced	partial
6	>=moderate	fully enforced	strong

Average weights

Attribute	Local	Global	Loc.norm.	Glob.norm.
ISN Development				
├─Environment and Technology	33	33	33	33
│ ├─Accessibility of material resources	71	24	71	24
│ │ ├─Availability of resources	43	10	43	10
│ │ └─Origin of resources	57	14	57	14
│ └─Proximity	29	10	29	10
│ ├─Distances	29	3	29	3
│ └─Installation/viability of heavy infrastructures	71	7	71	7
├─Organization and Economics	33	33	33	33
│ ├─Economy and economics	41	14	41	14
│ │ ├─Sustainable dynamics	50	7	50	7
│ │ └─Reduction of raw material and cost saving	50	7	50	7
│ ├─Learning organisation	41	14	41	14
│ │ ├─Participative management	57	8	57	8
│ │ └─Innovative ideas	43	6	43	6
│ └─Common IS platform	18	6	18	6
└─Social Forces	33	33	33	33
├─Institutions	33	11	33	11
│ ├─Urban planning	43	5	43	5
│ └─Regulations for waste management	57	6	57	6
├─Cognitive frames	33	11	33	11
└─Networks	33	11	33	11

Evaluation results

Attribute	Kalundborg	Aalborg	Guayama	Barceloneta	Kwinana	Gladstone	Rotterdam
ISN Development	fully developed	mostly developed	mostly developed	fully developed	mostly developed	fully developed	mostly developed
├─Environment and Technology	strong	strong	weak	strong	weak	strong	weak
│ ├─Accessibility of material resources	satisfying	satisfying	low	satisfying	low	satisfying	low
│ │ ├─Availability of resources	partially	partially	partially	yes	partially	partially	partially
│ │ └─Origin of resources	internal	internal	mixed	internal	mixed	internal	mixed
│ └─Proximity	moderate	moderate	moderate	moderate	moderate	moderate	moderate
│ ├─Distances	short	short	medium	medium	medium	short	short
│ └─Installation/viability of heavy infrastructures	incomplete	incomplete	incomplete	incomplete	incomplete	incomplete	incomplete
├─Organization and Economics	satisfying	moderate	satisfying	satisfying	satisfying	moderate	moderate
│ ├─Economy and economics	medium	medium	high	high	medium	medium	medium
│ │ ├─Sustainable dynamics	partial	partial	strong	strong	partial	partial	partial
│ │ └─Reduction of raw material and cost saving	partial	partial	partial	partial	partial	partial	partial
│ ├─Learning organisation	satisfying	satisfying	satisfying	satisfying	satisfying	satisfying	satisfying
│ │ ├─Participative management	partial	strong	partial	partial	partial	partial	partial
│ │ └─Innovative ideas	partial	partial	partial	strong	partial	partial	partial
│ └─Common IS platform	yes	partial	partial	yes	yes	partial	partial
└─Social Forces	satisfying	moderate	moderate	satisfying	moderate	satisfying	moderate
├─Institutions	strong	partial	partial	strong	strong	strong	partial
│ ├─Urban planning	satisfying	moderate	satisfying	satisfying	satisfying	satisfying	moderate
│ └─Regulations for waste management	fully enforced	partially enforced	partially enforced	fully enforced	fully enforced	fully enforced	partially enforced
├─Cognitive frames	strong	strong	partial	partial	partial	strong	partial
└─Networks	strong	partial	strong	strong	partial	partial	partial

List of figures

List of tables

References and sources

Aalborg Portland A/S. n.d. https://www.aalborgportland.dk/. Accessed on 10. December 2017.

Aalborg Portland Environmental Report. 2013. http://www.aalborgportland. com.cn/uploadfiles/report/2014-10/141387197585-ZsJOhA.pdf. Accessed on 10. May.

Aalborg Portland Miljøredegørelse 2011. http://www.aalborgportland.dk/media/ ap_miljoredegorelse2011.pdf n.d. Accessed on 13. May.

Aalborg-Renovation-Groenne-Regnskaber.Pdf. n.d. https://www.aalborg.dk/ media/6362280/aalborg-renovation-groenne-regnskaber.pdf. Accessed on 10. May.

Aalborg Supply. n.d. https://www.aalborgforsyning.dk/. Accessed on 10. May.

Abstractbook-Oral 050603.Pdf. n.d. http://is4ie.org/Resources/Documents/ Abstractbook-oral%20050603.pdf. Accessed on 10. May.

Adam, Frane, Matej Makarovič, Borut Rončević and Matevž Tomšič. 2005. *The challenges of sustained development: The role of socio-cultural factors in East-Central Europe.* New York; Budapest: Central European University Press.

AEBIOM. 2017. »DONG Energy Reaches Deal to Convert Asnæs Power Station to Biomass (from Bioenergy International).« *AEBIOM* (blog). 4 July 2017. http://www.aebiom.org/dong-energy-reaches-deal-to-convert-asnaes-power-station-to-biomass-from-bioenergy-international/. Accessed on 10. May.

Alkhalil, Adel. 2016. »A Model to Support the Decision Process for Migration to Cloud Computing.« 318. https://pdfs.semanticscholar.org/0de5/44e14a3aaea1 ecc742b4cf863ad1b5ad4b48.pdf. Accessed on 10. May 2017.

Allenby, Braden R. and William E. Cooper. 1994. »Understanding Industrial Ecology from a Biological Systems Perspective.« *Environmental Quality Management* 3 (3): 343–354. Accessed on 21. June. https://onlinelibrary.wiley. com/doi/abs/10.1002/tqem.3310030310

Andrea, Luciano D., Giancarlo Quaranta, and Gabriele Quinti. 2005. »Manuale Sui Processi Di Socializzazione Della Ricerca Scientifica e Tecnologica.« *CERFE. Rome.* Accessed on 22. May. http://www.cerfe.org/public/ ManualeRAST.pdf.

Archer, Margaret S. 1995. *Realist Social Theory: The Morphogenetic Approach.* Cambridge; New York; Melbourne: Cambridge University Press.

Archer, Margaret S. 2000. *Being Human: The Problem of Agency.* Cambridge; New York; Melbourne: Cambridge University Press.

Archer, Margaret S. 2003. *Structure, Agency and the Internal Conversation.* Economic, and Social Research Council (Great Britain); Cambridge; New York; Melbourne: Cambridge University Press. https://books.google.si/ books?id=KvK9O8R85KYC.

Archer, Margaret S. 2010. »Routine, Reflexivity, and Realism.« *Sociological Theory* 28 (3): 272–303.

Armstrong, Michael and Angela Baron. 1998. *Performance Management: The New Realities.* San Francisco, CA: State Mutual Book & Periodical Service.

Årsrapport-Aalborg Portland A/S. n.d. https://www.aalborgportland.dk/default. aspx?m=4&i=532&pi=1&pr=1. Accessed on 10. May.

Ashton, Weslynne. 2008. »Understanding the Organization of Industrial Ecosystems.« *Journal of Industrial Ecology* 12 (1): 34–51.

Ayres, Robert U. 1989. »Industrial Metabolism.« *Technology and Environment* : 23–49. https://books.google.si/books?id=7WMrAAAAYAAJ&pg=PA23&l pg=PA23&dq=Ayres,+Robert+U.+1989.+%C2%BBIndustrial+Metabolism .%C2%AB+Technology+and+Environment+:+23%E2%80%9349.&source =bl&ots=LCUl0aMtPX&sig=ACfU3U2T7i84dCqiBM8u811__LnExAm8M A&hl=sl&sa=X&ved=2ahUKEwiwjKizmsPkAhWBxIsKHbdfDrwQ6AEw AnoECAYQAQ#v=onepage&q=Ayres%2C%20Robert%20U.%201989.%20 %C2%BBIndustrial%20Metabolism.%C2%AB%20Technology%20and%20 Environment%20%3A%2023%E2%80%9349.&f=false. Accessed on 10. June.

Ayres, Robert U. and Leslie Ayres. 2002. *A Handbook of Industrial Ecology.* Elgar Original Reference Series. Edward Elgar Publishing, Incorporated. Accessed on 10, June. https://books.google.si/books?id=g1Kb-xizc1wC&prin tsec=frontcover&dq=https://books.google.si/books?id%3Dg1Kb-xizc1wC.& hl=sl&sa=X&ved=0ahUKEwje0fagm8PkAhXLlYsKHanGDrEQ6AEIKTAA #v=onepage&q&f=false.

Baas, Leenard W. 2005. *Cleaner Production and Industrial Ecology: Dynamic Aspects of the Introduction and Dissemination of New Concepts in Industrial Practice.* Netherland: Academische Uitgeverij Eburon, BD Utrecht.

Baas, Leenard W. and Frank Boons. 2007. »Industrial Symbiosis in a Social Science Perspective.« In *Industrial Symbiosis in Action. Report on the Third International Industrial Symbiosis Research Symposium Birmingham, England, August* edited by R. Lombardi and P. 5–6, 2006 (77–82). in, 77–82. England: Laybourn.

Baas, Lenard W. and Gijsberg Korevaar. 2010. *Eco-Industrial Parks in the Netherlands: The Rotterdam Harbor and Industry Complex. Sustainable Development in the Process Industries.* Hoboken, NJ: Wiley, 59–79.

Baas, Leo. 2005. *Cleaner Production and Industrial Ecology: Dynamic Aspects of the Introduction and Dissemination of New Concepts in Industrial Practice.* Eburon. https://books.google.si/books?id=QAVFuUi-uXUC&printsec=frontc over&hl=sl#v=onepage&q&f=false. Accessed on 24. June.

Baas, Leo. 2008. »Industrial Symbiosis in the Rotterdam Harbour and Industry Complex: Reflections on the Interconnection of the Techno-sphere with the Social System.« *Business Strategy and the Environment* 17 (5): 330–340.

Baas, Leo. 2011. »Planning and Uncovering Industrial Symbiosis: Comparing the Rotterdam and Östergötland Regions: Strategies for Manufacturing.« *Business Strategy and the Environment* 20 (7): 428–440. https://onlinelibrary. wiley.com/doi/abs/10.1002/bse.735

Baas, Leo and Don Huisingh. 2008. »The Synergistic Role of Embeddedness and Capabilities in Industrial Symbiosis: Illustration Based upon 12 Years of Experiences in the Rotterdam Harbour and Industry Complex.« *Progress in Industrial Ecology, An International Journal* 5 (5–6): 399–421.

Baas, Leo and Frank A. Boons. 2004. »An Industrial Ecology Project in Practice: Exploring the Boundaries of Decision-Making Levels in Regional Industrial Systems.« *Journal of Cleaner Production* 12 (8–10): 1073–1085.

Bard, Jonathan B. L. 1990. *Morphogenesis: The Cellular and Molecular Processes of Developmental Anatomy.* Cambridge, England: Cambridge University Press.

Barkow, Jerome H., Leda Cosmides, and John Tooby. 1995. *The Adapted Mind: Evolutionary Psychology and the Generation of Culture.* USA: Oxford University Press.

Beardwell, Ian and Len Holden. 1997. *HRM: A Contemporary Perspective.* London: Pitman.

Beckert, Jens. 2009. »The Social Order of Markets.« *Theory and Society* 38 (3): 245–269.

Beckert, Jens. 2010. »How Do Fields Change? The Interrelations of Institutions, Networks, and Cognition in the Dynamics of Markets.« *Organization Studies* 31 (5): 605–627.

Behne, Bradley A. 2016. Industrial Ecology Analysis of the Potential for an Eastern Nebraska Industrial Symbiosis Network (ENISN): A Comparative Study. MSc thesis. Lincoln: University of Nebraska. https://digitalcommons. unl.edu/cgi/viewcontent.cgi?article=1137&context=natresdiss. Accessed on 10. June 2017.

Bengtsson, Maria and Sören Kock. 2000. »«Coopetition« in Business Networks-to Cooperate and Compete Simultaneously.« *Industrial Marketing Management* 29 (5): 411–426.

Bhaskar, Roy. 1989. *Reclaiming Reality*. London: Verso.

Bhaskar, Roy. 2010. *Reclaiming Reality: A Critical Introduction to Contemporary Philosophy*. Taylor & Francis Routledge.

Bhidé, Amar V. 2003. *The Origin and Evolution of New Businesses*. New York: Oxford University Press, 244.

Black Sea Industrial Symbiosis Platform, European MSP Platform. 2017. http://msp-platform.eu/practices/black-sea-industrial-symbiosis-platform. Accessed on10. February.

Bohanec, Marko. 2003. »Decision Support.« In *Data Mining and Decision Support*, 23–35. Boston, MA: Springer.

Bohanec, Marko. 2008. *DEXi: Program for Multi-Attribute Decision Making User's Manual*. Ljubljana, Slovenia: Institut Jozef Stefan.

Bohanec, Marko, Venceslav Kapus, Bojan Leskošek, and Vladislav Rajkovič. 2000. *Talent: ekspertni sistem za usmerjanje otrok in mladine v športne panoge*. Ljubljana: Ministrstvo za šolstvo in šport: Zavod Republike Slovenije za šolstvo.

Bohanec, Marko and Vladislav Rajkovič. 1990. »DEX: An Expert System Shell for Decision Support.« *Sistemica* 1 (1): 145–157.

Bohanec, Marko, Blaž Zupan, and Vladislav Rajkovič. 2000. »Applications of Qualitative Multi-Attribute Decision Models in Health Care.« *International Journal of Medical Informatics* 58: 191–205.

Boons, Frank A. and Lenard W.Baas. 1995. »The Organisation of Industrial Ecology: The Importance of Regions.« *Proceedings of the 2nd European Roundtable on Cleaner Production*, Rotterdam, November 1–3, 1995.

Boons, Frank A. and Lenard W. Baas. 2006. »Industrial Symbiosis in a Social Science Perspective.« In *Discussion Proposal for the Third Industrial Symbiosis Research Symposium*, Birmingham, UK, August 5–6, 2006.

Boons, Frank and Jennifer A. Howard-Grenville. 2009. *The Social Embeddedness of Industrial Ecology*. UK: Edward Elgar Publishing.

Booth, Wayne C., Gregory G. Colomb, and Joseph M. Williams. 2009. *The Craft of Research, Third Edition*. University of Chicago Press, USA. https://books.google.si/books?id=Y31pUtkwb2oC&printsec=frontcover&hl=sl&source=gbs_ge_summary_r&cad=0#v=onepage&q&f=false. Accessed on 22. May.

Boshkoska, Biljana Mileva, Erika DžajićUršič, and Borut Rončević. 2015. »A Model for Evaluation of Appropriateness of Industrial Symbiosis Networking Based on DEX.« Upcoming paper-draft, not published. Internal document, on request.

Boshkoska, Biljana Mileva, Borut Rončević, and Erika DžajićUršič. 2018. »Modeling and Evaluation of the Possibilities of Forming a Regional

Industrial Symbiosis Networks.« *Social Sciences* 7 (1): 13. https://doi.
org/10.3390/socsci7010013.

Bossilkov, Albena, Rene van Berkel, and Glen Corder. 2005. »Regional Synergies
for Sustainable Resource Processing.« A Status Report.

Bourdieu, Pierre. 1986. The Forms of Capital (pp241258) in Handbook of
Theory and Research for the Sociology of Education. J. *Greenwood Press, New
York, USA.*

Callon, Michel. 1998. »Introduction: The Embeddedness of Economic Markets
in Economics.« *The Sociological Review* 46 (S1): 1–57.

Camisón, César and Beatriz Forés. 2011. »Knowledge Creation and Absorptive
Capacity: The Effect of Intra-District Shared Competences.« *Scandinavian
Journal of Management* 27 (1): 66–86.

Cavallo, Marino, Piergiorgio Degli Esposti, Kostas Konstantinou, Franko
Nemac, Matjaž Grmek, Tina Okorn, Neža Ana Rojko, et al. 2012. *Priročnik za
zeleno komuniciranje in marketing.* Ljubljana: ApE.

Chertow, Marian R. 2000. »Industrial Symbiosis: Literature and Taxonomy.«
Annual Review of Energy and the Environment 25 (1): 313–337. https://doi.
org/10.1146/annurev.energy.25.1.313.

Chertow, Marian R. 2004. *Economic and Environmental Impacts from
Industrial Symbiosis Exchanges: Guayama, Puerto Rico.* No. 0407. Rensselaer
Polytechnic Institute, Department of Economics.

Chertow, Marian R. 2004. *Encyclopedia of Energy.* Cleveland CJ: Editorial
Oxford.

Chertow, Marian R. Weslynne Ashton, and Juan Espinosa. 2008. »Industrial
Symbiosis in Puerto Rico: Environmentally Related Agglomeration
Economies.« *Regional Studies* 42 (December): 1299–1312. https://doi.
org/10.1080/00343400701874123.

Chertow, Marian R. 2007. »Multi-Scale Industrial Symbiosis.« *Journal of
Industrial Ecology* 11 (1): 11–30.

Chertow, Marian R. 2007. »«Uncovering« Industrial Symbiosis.« *Journal of
Industrial Ecology* 11 (1): 11–30. https://doi.org/10.1162/jiec.2007.1110.

Chertow, Marian R. and Weslynne S. Ashton. 2009. The Social Embeddedness
of Industrial Symbiosis Linkages in Puerto Rican Industrial Regions, In:
F.A. Boons, J. Howard-Grenville, eds., *The Social Embeddedness of Industrial
Ecology.* Edward Elgar Publishers, Cheltenham, UK,128–151.

Chertow, Marian R. and D. Rachel Lombardi. 2005. »Quantifying Economic
and Environmental Benefits of Co-Located Firms.« *Environmental Science &
Technology* 39 (17): 6535–6541. https://doi.org/10.1021/es050050+.

Chertow, Marian R. and D. Rachel Lombardi. n.d. »Exchanges: Guayama, Puerto Rico.« http://www.economics.rpi.edu/workingpapers/rpi0407.pdf. Accessed on 1. February 2018.

Christensen, Jørgen. 2004. »Presentation at Yale Industrial Symbiosis Research Symposium, January 9, 2004.« New Haven, CT.

Clemen, Robert T. and Robert L. Winkler. 1999. »Combining Probability Distributions from Experts in Risk Analysis.« *Risk Analysis* 19 (2): 187–203.

Cohen-Rosenthal, Edward and Judy Musnikow. 2017. *Eco-Industrial Strategies: Unleashing Synergy between Economic Development and the Environment.* Routledge, UK.

Coleman, James. 1994. Foundations of Social Theory. Harvard University Press, Cambridge, UK.

Corder, Glen. D. 2008. Developing local synergies in the Gladstone industrial area: Project 3C1. Centre for Social Responsibility in Mining, Sustainable Minerals Institute, The University of Queensland, Queensland, Australia.https://www.academia.edu/35723182/Potential_synergy_ opportunities_in_the_Gladstone_Industrial_Area. Accessed on 1. June 2016.

Corder, Glen D., Artem Golev, JulianFyfe, and Sarah King. 2014. »The Status of Industrial Ecology in Australia: Barriers and Enablers.« *Resources* 3 (2): 340–361.

Cosmides, Leda and John Tooby. 1992. »Cognitive Adaptations for Social Exchange.« *The Adapted Mind: Evolutionary Psychology and the Generation of Culture* 163: 163–228.

Costa, Inês and Paulo Ferrão. 2010. »A Case Study of Industrial Symbiosis Development Using a Middle-Out Approach.« *Journal of Cleaner Production* 18 (10–11): 984–992.

Costa, Inês, Guillaume Massard, and Abhishek Agarwal. 2010. »Waste Management Policies for Industrial Symbiosis Development: Case Studies in European Countries.« *Journal of Cleaner Production* 18 (8): 815–822.

CRESSI Publications Saïd Business School No. 5. 2015. CRESSI Publications Saïd Business School No. 5. 2015. Taking Action for Social Innovation.

http://eureka.sbs.ox.ac.uk/5950/1/CRESSI_Working_Paper_19_UK_Social_ Innovation_Policy_Edmiston.pdf. Accessed on 26. October 2017.

CTTÉI. 2013. Centre de Transfert Technologique en Écologie Industrielle, »Creating an Industrial Symbiosis«. http://gcpcenvis.nic.in/PDF/Industrial_ Symbiosis.pdf. Accessed on October 2017.

Czarniawska, Barbara and Bernward Joerges. 1996. »Travel of Ideas.« In *Translating Organizational Change* edited by Barbara Czarniawska-Joerges; Guje Sevón. Berlin; New York: Walter de Gruyter. De Gruyter studies in organization, 56. https://www.scribd.com/doc/176932296/CZARNIAWSKA-JOERGES-Translating-Organizational-Change-1996-Travel-of-Ideas

Damij, Nadja, Pavle Boškoski, Marko Bohanec, and Biljana M.Boshkoska. 2016. »Ranking of Business Process Simulation Software Tools with DEX/QQ Hierarchical Decision Model.« *PloS One* 11 (2): 6–16, e0148391. http://doi:10.1371/journal.pone.014839.

Déry, Richard, Chantale Mailhot, and Véronique Schaeffer. 2007. »13. Ethics and Management Education: The MBA under Attack.« *Moral Foundations of Management Knowledge*, 257–278.

Deutz, Pauline. 2014. »Food for Thought: Seeking the Essence of Industrial Symbiosis.« In R. Salomone and G. Saija (eds.), *Pathways to Environmental Sustainability: Methodologies and Experiences.* Switzerland: Springer. 3–11.

DEXi: A Program for Multi-Attribute Decision Making. http://kt.ijs.si/MarkoBohanec/dexi.html. Accessed on 6. January 2015.

DiMaggio, Paul J. and Walter W. Powell. 1983. »The Iron Cage Revisited: Institutional Isomorphism and Collective Rationality in Organizational Fields.« *American Sociological Review* 48 (2): 147–160.

DiMaggio, Paul J. and Walter W. Powell. 1991. *The New Institutionalism in Organisational Analysis.* USA: University of Chicago Press.

Diwekar, Urmila and Mitchell J. Small. 2002. »11. Process Analysis Approach to Industrial Ecology.« *A Handbook of Industrial Ecology*, 114. https://epdf.pub/a-handbook-of-industrial-ecology.html. Accesed on 8. January 2016.

Doménech A. Teresa. 2010. Social Aspects of Industrial Symbiosis Networks. PhD dissersatition. UCL (University College London).

Doménech A. Teresa and Michael Davies. 2011. »Structure and Morphology of Industrial Symbiosis Networks: The Case of Kalundborg.« *Procedia-Social and Behavioral Sciences* 10: 79–82.

Drucker, Daniel C. 1955. »The Effect of Shear on the Plastic Bending of Beams.« *Brown University Providence Ri.Div. of Engineering, Providence, USA.* https://apps.dtic.mil/dtic/tr/fulltext/u2/a950451.pdf. Accessed on 17. May 2016.

Duchin, Faye. 1998. *Structural Economics: Measuring Change in Technology, Lifestyles, and the Environment.* Washington, DC: Island Press.

Džajić Uršič Erika and Borut Rončević. 2017. »Industrial Symbiotic Networks in the Information Society: Research Challenges and Perspectives.« In *Information Society and Its Manifestations: Economy, Politics, Culture,* edited by Borut Rončević and Matevž Tomšič, 70–80, Bern, Switzerland: Peter Lang.

Edquist, Charles and Olle Edqvist. 1979. »Social Carriers of Techniques for Development.« *Journal of Peace Research* 16 (4): 313–331.

Efstathiou, Janet and Vladislav Rajkovič. 1979. »Multi-attribute Decision-Making Using a Fuzzy Heuristic Approach.« *IEEE Transactions on Systems, Man, and Cybernetics* 9 (6): 326–333.

Ehrenfeld, John and Nicholas Gertler. 1997. »Industrial Ecology in Practice: The Evolution of Interdependence at Kalundborg.« *Journal of Industrial Ecology* 1 (1): 67–79. https://doi.org/10.1162/jiec.1997.1.1.67.

Ehrenfeld, John R. 2000. »Industrial Ecology: Paradigm Shift or Normal Science?« *American Behavioral Scientist* 44 (2): 229–244.

Eilering, Janet A. M. and Walter J. V. Vermeulen. 2004. »Eco-Industrial Parks: Toward Industrial Symbiosis and Utility Sharing in Practice.« *Progress in Industrial Ecology, an International Journal* 1 (1–3): 245–270.

Esty, Daniel C. and Michael E. Porter. 1998. »Industrial Ecology and Competitiveness.« *Journal of Industrial Ecology* 2 (1): 35–43.

Fischhoff, Baruch and Mitchell J. Small. 1999. »Human Behavior in Industrial Ecology Modeling.« *Journal of Industrial Ecology* 3 (2–3): 4–7.

Fligstein, Neil. 2002. *The Architecture of Markets: An Economic Sociology of Twenty-First-Century Capitalist Societies.* New Jersey: Princeton University Press.

Fourcade, Marion. 2007. »Theories of Markets and Theories of Society.« *American Behavioral Scientist* 50 (8): 1015–1034.

Fric, Urška and Borut Rončević. 2018. »E-Simbioza-Leading the Way to a Circular Economy through Industrial Symbiosis in Slovenia.« *Socijalna ekologija: časopis za ekološku misao i sociologijska istraživanja okoline* 27(2): 119–140.

Frosch, Robert A. and Nicholas E. Gallopoulos. 1989. »Strategies for Manufacturing.« *Scientific American* 261 (3): 144–153. http://www.jstor.org/stable/24987406. Accessed on 6. November 2016.

Gallaud, Delphine and Blandine Laperche. 2016. *Circular Economy, Industrial Ecology and Short Supply Chain: Towards Sustainable Territories.* New Jersey: John Wiley & Sons.

Gertler, Nicholas. 1995. »Industry Ecosystems: Developing Sustainable Industrial Structures.« Thesis (M.S.). Massachusetts Institute of Technology, Institute of Technology, Dept. of Civil and Environmental Engineering, USA. http://hdl.handle.net/1721.1/11556. Accessed 18. April 2016.

Gibbs, David. 2003. »Trust and Networking in Inter-Firm Relations: The Case of Eco-Industrial Development.« *Local Economy* 18 (3): 222–236.

Gibbs, David and Pauline Deutz. 2005. »Implementing Industrial Ecology? Planning for Eco-Industrial Parks in the USA.« *Geoforum* 36 (4): 452–464.

Gibbs, David and Pauline Deutz. 2007. »Reflections on Implementing Industrial Ecology through Eco-Industrial Park Development.« *Journal of Cleaner Production* 15 (17): 1683–1695.

Giddens, Anthony. 1979. »Agency, Structure.« In *Central Problems in Social Theory*, 49–95. London, UK: Palgrave.

Gingrich, Craig. 2012. »Industrial Symbiosis: Current Understandings and Needed Ecology and Economics Influences.« *Ontario Center for Engineering and Publication*, 44–46. https://pdfs.semanticscholar.org/cbb 3/5fa904de20a1c6f713fe64a41291b806da96.pdf. Accessed on 4. February 2016.

Gnyawali, Devi R. and Ravindranath Madhavan. 2001. »Cooperative Networks and Competitive Dynamics: A Structural Embeddedness Perspective.« *Academy of Management Review* 26 (3): 431–445. https://doi.org/10.5465/ amr.2001.4845820.

Graedel, Thomas E. 1996. »On the Concept of Industrial Ecology.« *Annual Review of Energy and the Environment* 21(1): 69–98.

Graedel, Thomas E. and Braden R. Allenby. 2003. *Industrial Ecology*. New Jersey: Prentice Hall.

Granovetter, Mark. 1985. »Economic Action and Social Structure: The Problem of Embeddedness.« *American Journal of Sociology* 91 (3): 481–510.

Granovetter, Mark. 1992. »Economic Institutions as Social Constructions: A Framework for Analysis.« *Acta Sociologica* 35 (1): 3–11.

Granovetter, Mark S. 1977. »The Strength of Weak Ties1.« In *Social Networks*, edited by Samuel Leinhardt, 347–367. Academic Press. https://doi. org/10.1016/B978-0-12-442450-0.50025-0.

Harris, Steve and Colin Pritchard. 2004. »Industrial Ecology as a Learning Process in Business Strategy.« *Progress in Industrial Ecology, An International Journal* 1 (1/2/3): 89. https://doi.org/10.1504/PIE.2004.004673.

Healy, Kieran. 1998. »Conceptualising Constraint: Mouzelis, Archer and the Concept of Social Structure.« *Sociology* 32 (3): 509–522.

Heeres, R. R., Walter J. V. Vermeulen, and F. B. de Walle. 2004. »Eco-Industrial Park Initiatives in the USA and the Netherlands: First Lessons.« *Journal of Cleaner Production* 12 (8–10): 985–995.

Henrion, Max and Baruch Fischhoff. 1986. »Assessing Uncertainty in Physical Constants.« *American Journal of Physics* 54 (9): 791–798.

Hewes, Anne K. and Donald I. Lyons. 2008. »The Humanistic Side of Eco-Industrial Parks: Champions and the Role of Trust.« *Regional Studies* 42 (10): 1329–1342.

Howard-Grenville, Jennifer and Raymond Paquin. 2008. »Organizational
    Dynamics in Industrial Ecosystems: Insights from Organizational Theory.« In
    Changing Stocks, Flows and Behaviors in Industrial Ecosystems (Vol.1). Edward
    Elgar, Cheltenham, UK and Northampton, MA, 122–139. .

Industrial Ecology. n.d. »Some Directions for Research-Pre Publication Draft.«
    https://phe.rockefeller.edu/ie_agenda/. Accessed on 7. January.

Industrial, writer in. n.d. »Achieving Sustainable Development Through
    Industrial Ecology – Green Planet Ethics.« http://greenplanetethics.com/
    wordpress/achieving-sustainable-development-through-industrial-ecology/.
    Accessed 26 May.

Industry Involvement, CSBP. n.d. https://www.csbp.com.au. Accessed on 21. May.

Institute, Rocky Mountain. 1998. *Green Development: Integrating Ecology and
    Real Estate.* Vol. 9. New Jersey: John Wiley & Sons.

Jackson, Tim and Roland Clift. 1998. »Where's the Profit in Lndustrial Ecology?«
    Journal of Industrial Ecology 2 (1): 3–5. https://doi.org/10.1162/jiec.1998.2.1.3.

Jacobsen, Noel and Stefan Anderberg. 2005. »Understanding the Evolution
    of Industrial Symbiotic Networks: The Case of Kalundborg.« *Economics of
    Industrial Ecology: Materials, Structural Change, and Spatial Scales.* chapter
    11, 313–335.

Jacobsen, Noel B. 2006. »Industrial Symbiosis in Kalundborg, Denmark: A
    Quantitative Assessment of Economic and Environmental Aspects.« *Journal
    of Industrial Ecology* 10 (1–2): 239–255.

Jarvis, Barry. 2008. »Aalborg Industries.« *Energy Oil and Gas.* Blog, http://www.
    energy-oil-gas.com/2008/04/09/aalborg-industries/. Accessed on 9. April.

Jereb, Eva, Marko Bohanec, and Vladislav Rajkovič. 2003. *Dexi: Računalniški
    Program Za Večparametrsko Odločanje: Uporabniški Priročnik.* Moderna
    organizacija, University of Maribor, Slovenija.

Jereb, Eva, Uroš Rajkovič, and Vladislav Rajkovič. 2005. »A Hierarchical
    Multi-attribute System Approach to Personnel Selection.« *International
    Journal of Selection and Assessment* 13 (3): 198–205. https://doi.
    org/10.1111/j.1468-2389.2005.00315.x.

Johnson, Jeremiah, Shona Quinn, Emily Shelton and Zhizhou Zhang (2003),
    "Luchetti Industrial Park: Retrofitting an Industrial Park for Symbiosis."
    Research Paper for the Industrial Ecology Spring 2003 Course, Yale School
    of Forestry and Environmental Studies. New Haven, USA.https://slidex.
    tips/download/sustainable-industrial-development-model-for-puerto-rico.
    Accesed on 2. November.

Kalundborg Industrial Symbiosis, n.d. https://stateofgreen.com/en/profiles/
    kalundborg-municipality/solutions/kalundborg-industrial-symbiosis.
    Accessed on 12. May.

Kalundborg Symbiose, n.d. http://www.symbiosis.dk/en/. Accessed on 14. May 2017.

Kassinis, Georgios Ioannis. 1997. *Industrial Reorganization and Inter-Firm Networking: In Search of Environmental Co-Location Economies.* New Jersey: Princeton University.

Klein, R. L. 1995. *Methlie. LB: Knowledge-Based Decision Support Systems.* Chichester: Wiley.

Koenig, Andreas W. 2009. *Eco-Industrial Park Development; A Guide for North America.* 47. Peter Lowitt, AICP, of the Devens Enterprise Commission and Andreas Koenig, of Eco-Industry.org with assistance of the Environment, Natural Resources and Energy Division (ENRE) and Economic Development Division (EDD) of the American Planning Association. Massachusetts, USA.

Korhonen, Jouni and Juha-Pekka Snäkin. 2005. »Analysing the Evolution of Industrial Ecosystems: Concepts and Application.« *Ecological Economics* 52 (2): 169–186.

Krapež, Alenka and Vladislav Rajkovič. 2003. *Tehnologije Znanja Pri Predmetu Informatika: Vodnik Za Izpeljavo Sklopa Tehnologije Znanja.* Zavod Republike Slovenije za šolstvo, Ljubljana, Slovenija.

Kt.ijs. 2018. Bohanec, Marko. n.d. »A program for pretty drawing of trees« 4. https://kt.ijs.si/MarkoBohanec/pub/IS2007_DEXiTree.pdf. Accessed on 13. May 2017.

Kurup, Biji R. 2007. *Methodology for Capturing Environmental, Social and Economic Implications of Industrial Symbiosis in Heavy Industrial Areas.* Curtin University, Bentley, Australia.

Lambert, A. J. D. and Frank A. Boons. 2002. »Eco-Industrial Parks: Stimulating Sustainable Development in Mixed Industrial Parks.« *Technovation* 22 (8): 471–484.

Laybourn, Peter. 2015. »Industrial Symbiosis: Delivering Resource Efficiency and Green Growth.« *Birmingham: International Synergies.* http://www. International-Synergies. Com/Wp-Content/Uploads/2015/10/G7-Laybourn-International-Synergies.Pdf. Accessed on 26. October 2017.

Leigh, Michael and Xiaohong Li. 2015. »Industrial Ecology, Industrial Symbiosis and Supply Chain Environmental Sustainability: A Case Study of a Large UK Distributor.« *Journal of Cleaner Production,* Bridges for a More Sustainable Future: Joining Environmental Management for Sustainable Universities (EMSU) and the European Roundtable for Sustainable Consumption and Production (ERSCP) Conferences, 106 (November): 632–643. https://doi.org/10.1016/j.jclepro.2014.09.022.

Leoncini, Riccardo, Mauro Lombardi, and Sandro Montresor. 2009. »From Techno-Scientific Grammar to Organizational Syntax: New Production

Insights on the Nature of the Firm.« SSRN Scholarly Paper ID 1150055. Rochester: Social Science Research Network. https://papers.ssrn.com/abstract=1150055. Accessed on 4. March.

Letcher, Trevor M. and Daniel Vallero. 2011. *Waste: A Handbook for Management*. Academic Press, Cambridge, USA.

Levin, Irwin P. and Gary J. Gaeth. 1988. »How Consumers Are Affected by the Framing of Attribute Information before and after Consuming the Product.« *Journal of Consumer Research* 15 (3): 374–378.

Lifset, Reid. 1997. »A Metaphor, a Field, and a Journal.« *Journal of Industrial Ecology* 1 (1): 1–3. https://doi.org/10.1162/jiec.1997.1.1.1.

Lifset, Reid and Thomas E. Graedel. 2002. »Industrial Ecology: Goals and Definitions.« *A Handbook of Industrial Ecology*, 3–15. https://www.elgaronline.com/view/1840645067.00009.xml. Accessed on 16. February 2016.

McDermott, Gerald A. 2007. »Politics and the Evolution of Inter-Firm Networks: A Post Communist Lesson.« *Organization Studies* 28 (6): 885–908.

Melia, Michael. 2007. »Puerto Rican City Sees the Writing on the Wall.« *Los Angeles Times*.. http://articles.latimes.com/2007/dec/10/business/ft-puertorico10. Accessed on 10. December 2007.

Merz, Sinclair Knight. 2002. *Kwinana Industrial Area Economic Impact Study: An Example of Industry Interaction*. Sinclair Knight Merz, Sydney, Australia.

Milkovich, George T. and John W. Boudreau. 1997. *Personnel/Human Resource Management: A Diagnostic Approach*. Homewood: Richard Irwin.

Mirata, Murat and Tareq Emtairah. 2005. »Industrial Symbiosis Networks and the Contribution to Environmental Innovation: The Case of the Landskrona Industrial Symbiosis Programme.« *Journal of Cleaner Production* 13 (10–11): 993–1002.

Misra, Krishna B. 2008. *Handbook of Performability Engineering*. London: Springer, Berlin, Germany.

https://books.google.si/books?hl=sl&lr=&id=cPgXg3GIMAsC&oi=fnd&pg=PA1&dq=Misra,+Krishna+B.+2008.+Handbook+of+Performability+Engineering.+London:+Springer,+Berlin,+Germany.&ots=yeCjuFpuh1&sig=4_S2lNZ8thH358suk9udRnA7_HQ&redir_esc=y#v=onepage&q=Misra%2C%20Krishna%20B.%202008.%20Handbook%20of%20Performability%20Engineering.%20London%3A%20Springer%2C%20Berlin%2C%20Germany.&f=false. Accesed on 3. April.

Mizruchi, Mark S. 2007. »Political Economy and Network Analysis. An Untapped Convergence.« *Sociologica* 1 (2): 1–27.

Modic, Dolores and Borut Rončević. 2018. »Social Topography for Sustainable Innovation Policy: Putting Institutions, Social Networks and Cognitive Frames in Their Place.« *Comparative Sociology* 17 (1): 100–127.

Multiple Criteria Decision Making-International Society on MCDM. 2018. http://www.mcdmsociety.org/. Accessed on 17. May.

Müller, Vincent C. (ed.). 2013. *Philosophy and Theory of Artificial Intelligence* (Vol. 5) 60–75. Springer Science & Business Media, Berlin, Germany.

Noorman, Klaas Jan and Ton Schoot Uiterkamp. 2014. *Green Households: Domestic Consumers, the Environment and Sustainability*. Routledge, London, UK.

Pacheco, Natalia Criado, Carlos Carrascosa, Nardine Osman, and Vicente Julián Inglada. 2017. *Multi-Agent Systems and Agreement Technologies: 14th European Conference, EUMAS 2016, and 4th International Conference, AT 2016, Valencia, Spain, December 15-16, 2016, Revised Selected Papers*. Vol. 10207. Springer.

Pact-carbon-transition. 2010. Societal Dynamics of Energy Transition Final Research Report. http://www.pact-carbon-transition.org/delivrables/D-4.1.pdf. Accessed on 3. March.

Peddle, Michael T. 1993. »Planned Industrial and Commercial Developments in the United States: A Review of the History, Literature, and Empirical Evidence Regarding Industrial Parks and Research Parks.« *Economic Development Quarterly* 7 (1): 107–124.

Phillips, Paul S., Richard Barnes, Margaret P. Bates, and Thomas Coskeran. 2006. »A Critical Appraisal of an UK County Waste Minimisation Programme: The Requirement for Regional Facilitated Development of Industrial Symbiosis/Ecology.« *Resources, Conservation and Recycling* 46 (3): 242–264.

Porter, Michael E. 1998. *Clusters and the New Economics of Competition* (Vol. 76). Harvard Business Review Boston, USA.

Posch, Alfred. 2010. »Industrial Recycling Networks as Starting Points for Broader Sustainability-Oriented Cooperation?« Journal of Industrial Ecology 14 (2): 242–257.

Potts, Jason. 2000. *The New Evolutionary Microeconomics*. Edward Elgar Publishing, Cheltenham, UK.

Potts, Jason. 2001. »Knowledge and Markets.« *Journal of Evolutionary Economics* 11 (4): 413–431.

Preda, Alex. 2007. »The Sociological Approach to Financial Markets.« *Journal of Economic Surveys* 21 (3): 506–533.

Putnam, Robert D., Robert Leonardi, and Raffaella Y. Nanetti. 1994. *Making Democracy Work: Civic Traditions in Modern Italy*. Princeton University Press, New Jersey, USA.

Ricerca–azione Sulla Povertà e L`esclusione Sociale. 2015. http://doczz.it/doc/675160/ricerca%E2%80%93azione-sulla-povert%C3%A0-e-l-esclusione-sociale. Accessed on 8. May.

Rončević, Borut and Matej Makarovič. 2010. »Towards the strategies of modern societies: systems and social processes.« *Innovation–The European Journal of Social Science Research* 23 (3): 223–239.

Rončević, Borut and Matej Makarovič. 2011. »Societal steering in theoretical perspective: social becoming as an analytical solution.« *Polish Sociological Review* 176 (4): 461–472.

Rui, Jiali and Reinout Heijungs. 2010. »Industrial Ecosystems as a Social Network.« In *Knowledge Collaboration & Learning for Sustainable Innovation: 14th European Roundtable on Sustainable Consumption and Production (ERSCP) Conference and the 6th Environmental Management for Sustainable Universities (EMSU) Conference, Delft, The Netherlands, October 25–29, 2010.* Delft University of Technology; The Hague University of Applied Sciences; TNO.

Schiller, Frank, Alexandra S. Penn, and Lauren Basson. 2014. »Analyzing Networks in Industrial Ecology Review of Social-Material Network Analyses.« *Journal of Cleaner Production* 76: 1–11.

Schwarz, Sebastian. n.d. »Cities in Transition towards Zero Waste: A Case Study of Aalborg Municipality.« 67. https://projekter.aau.dk/projekter/en/studentthesis/byer-i-overgangen-til-zero-waste(f458ba6a-1bd9-4bd7-9787-929b89d88619).html. Accessed on 13. January.

Scott, John. 2000. *Social Network Analysis: A Handbook.* London: Sage. *2nd Edition.*

Scrase, James I., Andy Stirling, Frank W. Geels, Adrian Smith, and Patrick Van Zwanenberg. 2009. »Transformative Innovation: A Report to the Department for Environment, Food and Rural Affairs.« SPRU-Science and Technology Policy Research, University of Sussex. https://doi.org/10.13140/2.1.4629.5849.

Shi, Han, Marian Chertow, and Yuyan Song. 2010. »Developing Country Experience with Eco-Industrial Parks: A Case Study of the Tianjin Economic-Technological Development Area in China.« *Journal of Cleaner Production* 18 (3): 191–199.

Simsek, Zeki, Michael H. Lubatkin, and Steven W. Floyd. 2003. »Inter-Firm Networks and Entrepreneurial Behavior: A Structural Embeddedness Perspective.« *Journal of Management* 29 (3): 427–442.

Social Constructionism. 2018. http://groundedtheoryreview.com/2012/06/01/what-is-social-constructionism/. Accessed on 19. February.

Solomon, Susan, Intergovernmental Panel on Climate Change and Intergovernmental Panel on Climate Change, eds. 2007. *Climate Change*

2007: The Physical Science Basis: Contribution of Working Group I to the Fourth Assessment Report of the Intergovernmental Panel on Climate Change. Cambridge, New York: Cambridge University Press.

Spekkink, Wouter. 2016. *Industrial Symbiosis as a Social Process.* Developing theory and methods for longitudinal investigation of social dynamics and development of industrial symbiosis. Rotterdam: Erasmus Univerity research Rotterdam, Netherlands.

Statoil Scolded over Workers' Buyouts. 2017. http://www.newsinenglish. no/2017/03/30/statoil-scolded-over-workers-buyouts/. Accessed on 7. January.

Steward, Fred. 2008. *Breaking the Boundaries: Transformative Innovation for the Global Good* (Vol. 7). Nesta, London, UK.

Suh, Sangwon, ed. 2009. *Handbook of Input-Output Economics in Industrial Ecology* (Vol. 23). Springer Science & Business Media, Berlin, Germany.

Synergies-Kwinana Industries Council. n.d. https://www.kic.org.au/ environment/synergies.html. Accessed on 18. March.

Sztompka, Piotr. 1999. *Trust: A Sociological Theory.* Cambridge University Press, USA.

Thongplew, Natapol. 2015. *Greening Production and Consumption: The Case of the Appliance and Dairy Industries in Thailand.* Wageningen: Wageningen University.

Thongplew, Natapol, Gert Spaargaren, and C. S. A. Van Koppen. 2014. »Greening Consumption at the Retail Outlet: The Case of the Thai Appliance Industry.« *International Journal of Sustainable Development & World Ecology* 21 (2): 99–110.

Tsoukas, Haridimos. 1991. »The Missing Link: A Transformational View of Metaphors in Organizational Science.« *Academy of Management Review* 16 (3): 566–585.

Tudor, Terry, Emma Adam, and Margaret Bates. 2007. »Drivers and Limitations for the Successful Development and Functioning of EIPs (Eco-Industrial Parks): A Literature Review.« *Ecological Economics* 61 (2–3): 199–207.

Uzzi, Brian. 1996. »The Sources and Consequences of Embeddedness for the Economic Performance of Organizations: The Network Effect.« *American Sociological Review* 61 (4): 674–698.

Uzzi, Brian. 1997. »Social Structure and Competition in Interfirm Networks: The Paradox of Embeddedness.« *Administrative Science Quarterly* 42 (1): 35–67.

Van Beers, Dick, Glen David Corder, Albena Bossilkov, and Rene Van Berkel. 2007a. »Regional Synergies in the Australian Minerals Industry: Case-Studies and Enabling Tools.« *Minerals Engineering* 20 (9): 830–841.

Van Beers, Dick, Albena Bossilkov, Glen Corder and Rene van Berkel. 2007b. »Industrial Symbiosis in the Australian Minerals Industry: The Cases of Kwinana and Gladstone.« *Journal of Industrial Ecology* 11 (1): 55–72. Accessed 14 January. https://doi.org/10.1162/jiec.2007.1161.

Van den Bergh, Jeroen C. J. M. and Marco Janssen. 2004. *Economics of Industrial Ecology: Materials, Structural Change, and Spatial Scales.* MIT Press, Cambridge, USA.

Van den Bosch, Suzanne. 2010. *Transition Experiments: Exploring Societal Changes towards Sustainability.*

Van Koppen,C.S.A Kris and Arthur P.J., 2002. Ecological Modernization of Industrial Ecosystems. In Lens, P., L.W.Hushoff Pol, P.Wilderer and T.Asano (eds.). Water Recycling and Resource Recovery in Industry: Analysis, Technologies and Implementation. IWA Publishing. https://www.academia.edu/26109798/The_Ecological_Modernisation_Reader_Environmental_Reform_in_Theory_and_Practice. Accessed on 1. May.

Van Merriënboer, Jeroen J. G. 1997. *Training Complex Cognitive Skills: A Four-Component Instructional Design Model for Technical Training.* Educational Technology, Publications Engelwood Cliffs, New Jersey, USA.

Velenturf, Anne P. M. and Paul D. Jensen. 2016. »Promoting Industrial Symbiosis: Using the Concept of Proximity to Explore Social Network Development.« *Journal of Industrial Ecology* 20 (4): 700–709.

Wallner, Heinz Peter. 1999. »Towards Sustainable Development of Industry: Networking, Complexity and Eco-Clusters.« *Journal of Cleaner Production* 7 (1): 49–58. https://doi.org/10.1016/S0959-6526(98)00036-5.

Wallner, Heinz P. and Michael Narodoslawsky. 1996. »Evolution of Regional Socio-Economic« Systems toward »Islands of Sustainability«.« *Journal of Environmental Systems* 24: 221–240.

Wallner, Heinz P., Michael Narodoslawsky, and Franz Moser. 1996. »Islands of Sustainability: A Bottom-Up Approach towards Sustainable Development.« *Environment and Planning A: Economy and Space* 28 (10): 1763–1778.

Wasserman, Stanley and Katherine Faust. 1994. *Social Network Analysis: Methods and Applications.* Cambridge: Cambridge University Press. https://doi.org/10.1017/CBO9780511815478.

Watts, Duncan J. 1999. *Small Worlds: The Dynamics of Networks Between Order and Randomness.* Princeton Studies in Complexity. Princeton University Press and Oxford, UK. https://books.google.si/books?hl=sl&lr=&id=soCe7RulvZcC&oi=fnd&pg=PP15&dq=Watts,+Duncan+J.+1999.+Small+Worlds:+The+Dynamics+of+Networks+Between+Order+and+Randomness.+Princeton+Studies+in+Complexity.+Princeton+University+Pre

ss.+&ots=6TdJassTzm&sig=a_Uz6i3BI-frXwmrNktnwrVNAuA&redir_
esc=y#v=onepage&q=Watts%2C%20Duncan%20J.%201999.%20Small%20
Worlds%3A%20The%20Dynamics%20of%20Networks%20Between%20
Order%20and%20Randomness.%20Princeton%20Studies%20in%20
Complexity.%20Princeton%20University%20Press.&f=false. Accesed on 2.
April 2017.

Watts, Duncan J. 2004. *Six Degrees: The Science of a Connected Age.* W. W. Norton
Company, NY, London. https://books.google.si/books?id=1gueFWR7qjoC&pri
ntsec=frontcover&dq=Watts,+Duncan+J.+2004.+Six+Degrees:+The+Science+
of+a+Connected+Age.+W.+W.+Norton.&hl=sl&sa=X&ved=0ahUKEwjcm5S_
scrkAhVM-qQKHY92DeMQ6AEIKDAA#v=onepage&q=Watts%2C%20
Duncan%20J.%202004.%20Six%20Degrees%3A%20The%20Science%20of%20
a%20Connected%20Age.%20W.%20W.%20Norton.&f=false. Accesed on 3.
April 2017.

Weber, Alfred, and Carl J. Friedrich. 1929. Alfred Weber's theory of the location
of industries. Chicago, Ill: The University of Chicago Press, USA.

Wright, Ramsey A., Raymond P. Côté, Jack Duffy, and John Brazner. 2009.
»Diversity and Connectance in an Industrial Context.« *Journal of Industrial
Ecology* 13 (4): 551–564.

Zhu, Qinghua, Ernest A. Lowe, Yuan-an Wei, and Donald Barnes. 2007.
»Industrial Symbiosis in China: A Case Study of the Guitang Group.« *Journal
of Industrial Ecology* 11 (1): 31–42.